Natural Gas Networks Performance After Partial Deregulation

Five Quantitative Studies

World Scientific Series on Energy and Resource Economics

(ISSN: 1793-4184)

Published

Vol. 1 Quantitative and Empirical Analysis of Energy Markets
by Apostolos Serletis

Vol. 2 The Political Economy of World Energy: An Introductory Textbook
by Ferdinand E Banks

Vol. 4 Energy, Resources, and the Long-Term Future
by John Scales Avery

Vol. 5 Natural Gas Networks Performance after Partial Deregulation: Five Quantitative Studies
by Paul MacAvoy, Vadim Marmer, Nickolay Moshkin & Dmitry Shapiro

Forthcoming

Vol. 3 Bridges Over Water: Understanding Transboundary Water Conflict, Negotiation and Cooperation
by Ariel Dinar, Shlomi Dinar, Stephen McCaffrey & Daene McKinney

World Scientific Series on Energy and Resource Economics – Vol. 5

Natural Gas Networks Performance After Partial Deregulation

Five Quantitative Studies

Paul MacAvoy
Yale University, USA

Vadim Marmer
University of British Columbia, Canada

Nickolay Moshkin
Cornerstone Research, USA

Dmitry Shapiro
University of North Carolina Charlotte, USA

NEW JERSEY • LONDON • SINGAPORE • BEIJING • SHANGHAI • HONG KONG • TAIPEI • CHENNAI

Published by

World Scientific Publishing Co. Pte. Ltd.
5 Toh Tuck Link, Singapore 596224
USA office: 27 Warren Street, Suite 401-402, Hackensack, NJ 07601
UK office: 57 Shelton Street, Covent Garden, London WC2H 9HE

British Library Cataloguing-in-Publication Data
A catalogue record for this book is available from the British Library.

World Scientific Series on Energy and Resource Economics — Vol. 5
NATURAL GAS NETWORKS PERFORMANCE AFTER PARTIAL DEREGULATION
Five Quantitative Studies

ISBN-13 978-981-270-860-1
ISBN-10 981-270-860-X

Typeset by Stallion Press
Email: enquiries@stallionpress.com

Printed in Singapore.

PREFACE

As the first word on a set of quantitative studies of natural gas markets, the preface should indicate who it is intended for, what those readers will find, and what they should take away. All of that is provided in the following five chapters in the context of quantitative studies that intend to be current and topical on certain energy markets, policy and their effects on the economy.

This book contains five semi-independent analyses of aspects in the performance of natural gas markets in Canada and in the United States in the recent past decade. The technology of natural gas, interacting with the economics of costs and demands for services to deliver gas to the consumer's burner tip, determines the first level of industry activity in inground gas discovery, development, and production. The second level, at the wellhead, consists of gathering, refining, and transfer of the product to pipeline nodes for transport to other intersecting nodes and ultimately to the third level, that of the distribution system at the point of use. At each point of transfer, there are buy-sell markets and also state and national agencies to regulate the buy-sell process. We carry out analyses of the response of service suppliers at these levels to changes in the extent of regulation in the early 1990s. Our results are intended to inform readers concerned with the effects on gas prices at various points in the networks as a result of such changes known as regulatory "reform", or "partial deregulation".

We do not account for the effects of all such changes inherent in "partial deregulation" in the last decade and a half. Instead, we concentrate on what we consider to be the building blocks for changing the performance of the supply side of sales markets. Decisions of the Federal Energy Regulatory Commission (FERC) required the separation of gas ownership from pipeline transportation services, the development of exchange markets for gas and (separate) transportation at hubs that were between field intake and city gate delivery. FERC also required the elimination of price controls on all transactions in product or line space except on firm contracts for delivery space. We seek the "facts" on changes in the geographic dimensions of the markets for line space, and on the extent of storage either at the wellhead or in reservoirs close to final sales locations. The focus, however, is on what happens to gas prices, both in the year-to-year markets and in the long run relative to what would have happened without this partial deregulation.

Our intention is to provide the reader with "economic knowledge" on how in supply/demand equilibrium the representative gas price changed because of modifications in the regulatory constraints on supply. This calls for testing models of price behavior in monopoly or oligopoly (supplier interactive) transportation markets with data on prices and volumes. The readers will encounter summary statistics for voluminous data on gas shipped, basis differentials in spot prices on sales at different hub locations, and pipeline charges, costs and earnings. These data are used to test hypotheses that pipelines increased their interaction with other lines, and increased short term and season-to-season space to meet volatile demands for services after partial deregulation. These hypotheses have been tested using various statistical measures, commonly referred to as "low brow" econometrics. The reader will encounter attempts at rigor in testing, of the form of co-integration techniques to define markets, and regression analysis to distinguish regulated monopoly service provision from partially deregulated Bertrand oligopoly service provision. The emphasis throughout is on tests to support a statement that gas sales and transportation have become less monopolistic in character, with prices lower but more volatile over the summer to winter heating seasons.

The quantitative studies cover some of the major impacts of FERC deregulation initiatives of the 1990s. There were earlier initiatives as a result of the Natural Gas Policy Act of 1978 which phased out FERC price controls of wellhead gas. The impacts of this earlier decontrol carried over into the later period (particularly, as will be seen, in the behavior of storage markets). And the 1990s deregulation also affected the performance of large industrial gas consumers, and local distributors at retail, on the buying side. We do not undertake quantitative studies of all these effects. The analyses here of price-cost margins and profitability of the major pipelines only allow us to state stylized facts about transportation and gas sales markets becoming more oligopolistic and less profitable.

What is encountered here, however, is more than what is found in the energy trade magazines on two dimensions. It is statements about the "type" of prices and services likely to be found in current markets — prices that have been driven below levels that would have been quoted by a single pipeline subject to 1950s style regulation. This lower level of prices is sustainable only if it generates margins over fixed and variable costs sufficient to lead to capacity expansion necessary to meet future increased demands. The magazines do not discuss, never mind document, such matters; their focus is on week to week pipeline space excesses or shortages, gas price spikes at retail, and so forth. Thus, the reader should take away unique "facts" on market definition, oligopoly pricing, capacity expansion, long run supply-demand equilibrium.

The semi-independent studies here were undertaken at the Yale University School of Management and Economics department by Professor Paul MacAvoy and PhD students Vadim Marmer, Nickolay Moshkin, and Dmitry Shapiro. Research grants necessary to provide the financial resources for quantitative work were provided by the John M. Olin Foundation. Graduate students in the MBA program, as Olin Fellows, compiled the databases; in particular Gene Agashin and Carlos Reyes set up the methodology to compile statistics for all major pipelines; Fellows Gene Agashin, Rumundaka Winodi, and Marni Rapaport set up and/or compiled price data for the basis differentials analysis; and Fellow Reyes with Winodi developed the data foundations for the analyses of pipeline financial performance over the 1990s. The chapter studies themselves were undertaken by the authors so designated. This does establish independency for each study and to a certain extent the value attached to authorship goes to those named; but all those on the title page read and edited all the pages in this book and the results of repetitive critiques by all four of us can be found on each chapter. We dedicate this work to the culture of interactive scholarship that marks economics and management education at Yale University. And we thank Jessica Lather Washburn of the Technical Support Staff of Mt. Ascutney Hospital and Health Center for turning the pages into a finished book manuscript.

Paul MacAvoy

CHAPTER 1

Introduction

Paul W. MacAvoy

Of the infrastructure network industries in the United States, surface transportation has proceeded the greatest distance from public utility-style price regulation to day-to-day market-determined pricing. Railroad, airline, and trucking services were subjected to regulatory reform in thc 1980s in which price controls administered by the Interstate Commerce Commission and Civil Aeronautics Board by and large were eliminated by congressional legislation, as were these agencies themselves. But regulatory agency controls of pricing and contract service offerings of the natural gas pipeline companies, while proceeding down the deregulatory pathway did not make it the 90 percent plus to open markets of the others.

Regulation in the early 1980s, had the three dozen large pipelines buying gas at the wellhead or separating plant from oil and gas producers in short and long term contracts, to be delivered in a merchant gas plus transportation package service to line hubs outside of the major population centers for gas industrial use or retail distribution by the local gas utilities. The links between entry hubs and delivery hubs at which wholesale transactions took place in the package traversed the country, at first from the older Texas and Louisiana fields to Boston, Chicago, Minneapolis, San Francisco, and then from new gas fields in Western Canada or the Rocky Mountains to these same exit hubs. At that point, the field price had been decontrolled under the Natural Gas Policy Act of 1978; and bid prices were on gas for long term contracts, and were rolled into the regulated transportation charge to comprise a merchant price at the exit hub. The charge was based on the historical costs of building the pipeline, plus costs of capital and operations, split between a reservation charge for space and a delivery charge for gas at the end of the link. The links were built at different times, and new fields were developed as demand expanded, so there was extensive overlap of the networks of different companies with different gas and delivery charges.

Looking at a national pipeline map, one would not likely discover large differences between hub-link-hub layouts and those for the airlines, or even the trucking companies. At the major cities, there were delivery terminals — and transfer terminals — for more than two or three service suppliers less than seven or eight such suppliers. This number and the rate of entry of potential suppliers made a case for deregulation — for prices not set by agency review of historical costs but in competitive interplay of numerous sources of service. That case was not made in the congressional reviews that set out the reform proposals for gas transportation. The agency itself — the Federal Energy Regulatory Commission — through public notice and hearings produced a rulemaking (Order 636 in 1992) that constituted partial deregulation.

There were more than a dozen major changes required in the transactions, ownership and interactions of service suppliers in Order 636. The pipeline was required to separate its purchases/sales of gas from its provision of transportation, and to treat an exchange of transport with others on an equal-service basis. All shippers were to be provided with offers of firm, interruptible, and secondary market (space already leased to others) service on an equal basis. This service array was to be provided at "market hubs", where pipelines overlapped, on electronic bulletin boards which also would contain the data of spot market prices and quantities of natural gas.

But pipeline transportation rates were not fully decontrolled, as were airline fares or trucking rates. In "bidweek" auctions, space offerings of lines were put up for lease. The rates for "firm" service that committed space ahead for various periods were to be regulated on a straight fixed variable basis, the "fixed" charge for reserving space was to cover historical plant costs and the "variable" charge was to equal operating costs on the line. On interruptible sales of space, the bidweek (variable) price could not exceed the average of these two regulated rates. On secondary sales, where a shipper puts its firm contract space back on the market, this new system let the market price prevail.

Other parts of Order 636 had to do with release of the newly liberated pipelines from public utility style obligations to serve as a supplier of last resort, or from requirements for non-discrimination in receiving or passing on gas to other pipelines at intermediate hubs.

This FERC order was a new blend of controls with market discipline, in a new set of pricing rules. It called for the agency to set specific prices, rather than to limit earnings to historical fixed costs. If there were ten transactions, only the two part prices on the two new contracts for commitment of space ahead were required to be limited, or capped, on the average cost per unit space. All the other eight space commitments, made on an interruptible or second provider basis, were not subject to that cap but were at higher or lower prices set by market conditions.

The intention was to change market structures, not set new pricing rules, so that the market would set prices competitively. In the Commission's words, "this rule requires significant alterations in the structure of interstate pipeline services…[and] will therefore reflect and finally complete the evolution to competition…so that all natural gas suppliers, including the pipeline as merchant, will compete for gas purchasers on an equal footing[1]". What was before the rulemaking a single company with multiplicity of lines, each with its own set of wellhead contracts for a gas volume for transport to a separate city gate, was to become a collection of lines from wellheads to a hub where there were other companies with duplicative lines. Unbundling was to break down integrated node-link systems into segments with duplicate links between shared nodes. That would replace one buyer, the pipeline, at the wellhead with multiple buyers (if not at that wellhead, then at intermediate nodes).

See the illustration in Fig. 1.1. Nodes in the first pipeline network consist of field entry node A and city gate exit node D; nodes in the second pipeline are field entry node

[1]FERC [Docket RM91-11-000] 56PR38372, pp. 7–8

Figure 1.1.

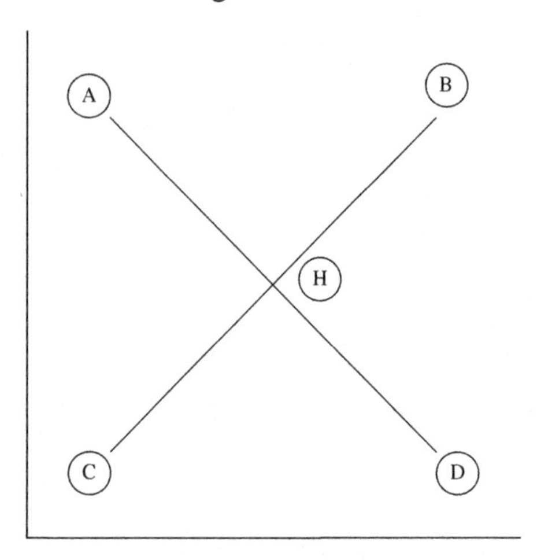

B and city gate exit C. Before Order 636, lines AD and BC were the only buyers in their respective wellhead markets, and hub *H* was no more than an intersection point of the two exclusive links; after 636, the agency's intention was that BC, or brokers with space on BC, could buy gas at *A*, have it transported to *H* on AD, then transfer it at *H* for delivery at *C*. There would be a multiplicity of gas buyers at each entry node, and for space to the nearest gathering hub. There also would be a multiplicity of sellers at exit nodes given open access to space from the last exchange hub.

With more than a dozen years of post-636 experience, the basic question is whether the performance of the gas transportation industry has changed to produce the results that would constitute such "structural reform". Has the replacement of price regulated merchant gas plus transportation on long term contracts with this fragmented spot markets for line space made price behavior more competitive?

This book seeks to provide an analytical response to that question. While the question itself barely takes a line, that response takes volumes because market structure, changed as indicated, and prices set in a new complex of markets, have added numerous strategic dimensions to pipeline behavior. It has been necessary to radically change market definitions; certainly the well-head gas entry markets for line space have broadened as a result of unbundling. We have attempted in Chapter 2, to determine whether there has been in the past order 636 period the emergence of one national market, in which all separate entry hubs are alternatives, or of a set of many regional markets. It has proved to be critical for analysis further on of pricing to determine whether the service provision of the Northeastern pipelines has become integrated with those of the Midwest and West coast pipelines. The determination to the best of our abilities has been intensely empirical.

Within these regional markets, the spot gas prices on the electronic bulletin boards have proved to be a new source of key information on the competitiveness of transportation charges. We have undertaken a project to document differences between entry and exit hub spot prices ("basis differentials"), to provide the foundation for an inference as to which types of oligopoly strategies have prevailed between these pipelines. Simple inspection indicates that in numerous entry hub locations, and almost all exit hub locations, three or four of the large incumbent pipeline network systems after order 636 provided alternative line space access. Perfect competition with line prices equal to marginal costs cannot prevail. But then there is the critical issue of whether these separate service corporations have been able to collectively control bidweek transport price setting in cartel-like fashion (with stable shares), or have at the other extreme suffered both off peak price collapses (along with on peak regulated price caps) of the type designated Bertrand Oligopoly. Nickolay Moshkin in Chapter 3 develops new econometric procedures for determining an answer based on the first few post-636 years experience.

This new pricing information, from new types of transactions, on its face does not appear to be different from the earlier pricing results of regulated transactions. From 1980 to the early 1990s as markets adjusted to phased deregulation of wellhead prices on long term contracts from the Natural Gas Policy Act of 1978, wellhead prices in constant dollars increased from \$1.00 mBtu (1978) to \$2.50 mBtu (1983) and then went down to slightly more than

\$1.00 mBtu (in 1990–1991).[2] From the mid-1990s to 2003, the first phase of partial deregulation of gas transportation under FERC Order 636, wellhead prices for spot gas increased from the \$1.50 mBtu level to \$6.50 mBtu in current dollars, or to half that in comparable constant dollars.[3] After declining to \$2.25, the spot average price increased to in excess of \$4.00 mBtu. In common, there have been extremely large increases and decreases in the wellhead price, even with basic changes in the structure of market transactions.

More penetrating analysis is required, centered on the performance of those companies subject to new deregulation requirements for common carriage service, and to divestiture of their gas contracts, and for specific caps on firm contracts for line space. To begin, we ask whether the major incumbent pipelines have fared well post-636; if Order 636 intended to make transportation more competitive then post-636 profitability should have been less, on the assumption that less-than-perfect pre-636 regulation failed to control the pipelines in ways that achieved elimination of monopoly profits.

---

[2] As in Fig. 1.2 of MacAvoy, P.W. (2000). *The Natural Gas Market — Sixty Years of Regulation and Deregulation*, Yale University Press, New Haven, p. 6.

[3] See sources for the analysis of price spikes in Chapter 4.

Our preliminary test is to estimate rates of return on pipeline investments, exclusive of those returns required to recover previous investments and to compensate continuing investors for the use of these funds. The excess of the realized rate of return over these required rates comprises EVA, the "economic value added" or excess profitability of post-636 operations. Based on the reported financial statements to the Security and Exchange Commission and Form Two Reports on gas financials of 31 pipelines, most generated no EVA through the 1990s. The exception was the pipeline systems in the Midwest, with between 2 and 5 percent returns in excess of the costs of capital from year to year over the period from 1992 to 1999. Pipeline systems on the East and West coast in some years realized 1 percent, but in other years were from 3 to 4 percent negative EVA. All the pipeline companies in the 31 observation samples were short returns necessary to cover the costs of capital in the early 2000s, when wellhead gas prices doubled. Again, there seems to be no pattern in keeping with the hypothesis that partial deregulation targeted at the pipelines had a pro-competitive effect on their earning performance. In the more recent years, that performance most likely fell short of what could be expected from long-term operations in competitive, not to mention monopoly markets.

Figure 1.2. EVA (Selected Companies) Using Market Values for WACC Calculation

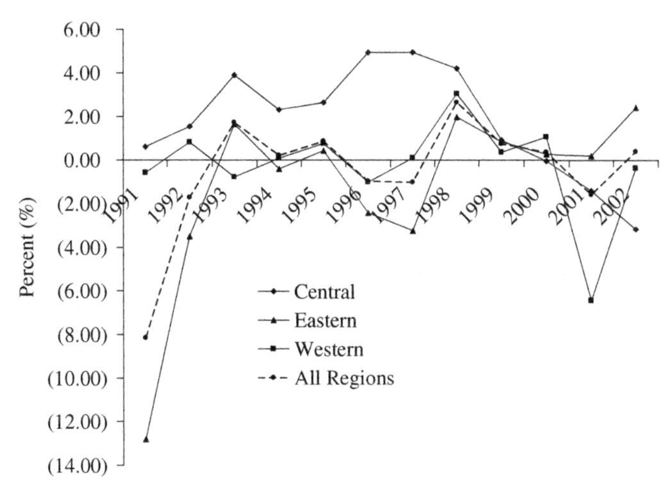

In the following chapters, we undertake more specific tests of performance designed to provide the basis for assessment of the effects of FERC deregulation initiatives. Chapter 2, based on research undertaken for this project by Vadim Marmer and Dmitry Shapiro, while they were doctoral candidates in Yale's Economics Department, seeks to set sub-national boundaries on markets for pipelines space by utilizing econometric measures of the extent of co-integration. Chapter 3 develops an econometric model of these markets to estimate parameters required for an assessment of whether the post-636 pipeline pricing was characterized by cartel, Cournot, or Bertrand behavior (a revised chapter from Nickolay Moshkin's Yale Doctoral dissertation). Chapter 4 describes the large and relatively frequent spikes in the basis differentials between spot gas entry and exit hubs on these pipelines, a phenomenon unique to the Eastern and Western sub-national pipeline service markets; this research, developed by Paul MacAvoy with the Olin Fellows in the MBA program at the Yale School of Management, indicates significant shortfall in pipeline capacity in the Eastern and West markets, with price controls passing on to gas brokers the profit gains from the spikes. Chapter 5 contains an analysis of shifts of storage volumes after Order 636 towards the producing region for reasons that are surprising at first, but after consideration of secondary effects from earlier Congressional initiatives are quite consistent with the movement to reduce regulation. The last chapter raises questions as to the long run effects on gas supply and demand, and with a national econometric model of MacAvoy's updated by Vadim Marmer it is found that the post-636 behavior on both sides of the (model) market are more responsive to outside impacts from oil prices.

CHAPTER 2

Quantitative Study Number One

REGIONAL MARKETS FOR GAS TRANSMISSION SERVICES

Vadim Marmer and Dmitry Shapiro

2.1. Introduction

Natural gas transmission is regulated by the Federal Regulatory Commission (FERC), at the margin, in terms of price limits on transportation services at peak demand periods (MacAvoy, 2000). The FERC determined limits on prices for shipment of gas at peak demand periods prevent the pipelines from realizing higher returns from higher bids for space in the pipelines. Instead, additional revenues generated from higher gas prices on sales to local distributors go to gas brokers with rights to the gas received at the delivery node. Such a situation may undermine pipelines' incentives to invest in construction of new lines and expansion of existing ones, to create a situation in which the existing pipeline capacity is not expanding to meet demands on transmission at the regulated peak prices. Even though profit gains, given the limits on prices, might be insufficient to finance the construction of new pipelines, new storage facilities might be able to relieve the peak-period excess demand by taking previously shipped supplies out of facilities near the arrival city gate. However, partial deregulation distorts and considerably delays construction of new storage facilities as well. Both sets of distortions determine new limits of "the market" for gas at receiving nodes.

In this chapter, we use a simple econometric procedure that allows identification of geographical limits on regional markets resulting from partial deregulation. Our analysis is based on the following premise. We assume that, in a network comprising "a market" temporary gas price shocks at one local receiving node should have an impact on spot prices at other nodes in the network as well which die out fast because of arbitrage. As a result, in the long-run, prices at every node in "the market" follow the pattern of the law of one price[1]. On the other hand, in the case of a separation of markets of two nodes of the network, their spot prices demonstrate diverging patterns for prolonged periods of time.

In order to illustrate this point, consider the following simple example. Assume that the network consists of only three hubs, one gas entry hub (Henry Hub) and two receiving hubs (in Chicago and New York City). Consider a positive demand shock in Chicago, caused by an extended period of extremely cold weather that does not reach to the eastern seaboard. Such a shock will lead to an increase in spot prices at the Chicago hub, making Chicago more attractive for spot sales by producers and gas brokers relative to New York City hub sales. In the absence of capacity constraints on gas flows, more throughput would be directed to the Chicago hub and less to New York City[2], which will lead to decreases in spot prices in Chicago and increases in New York City. Moreover, price adjustments should occur no matter whether the two hubs

---

[1]In equilibrium, the difference in prices between any two nodes should be approximately equal to transportation cost. Clearly this assumption is too strong, however, for our analysis we need only a much weaker assumption of econometric stationarity of other factors.

[2]The exception is the case when the supply of gas is perfectly inelastic with respect to contract delivery point.

supply is perfectly competitive, or served by a pipeline with a monopoly. There is an exception that can prevent the two locations from being in the same market. Assume first that the pipeline links to Chicago are fully "booked" with contracts for space with gas suppliers. Even though Chicago has become more attractive as a delivery point, producers at Henry Hub cannot take advantage of the difference in the spot prices to ship more to that location. As the consequence of such a link capacity constraint, the spot prices in Chicago increase, while prices in New York City remain without any change. That is, a local demand shock in a partially blocked network does not lead to an increase in prices over the network; the net link with specific capacity limits experiences local bottlenecks that cause sharp price differences between just two local receiving hubs.

Our analysis consists of two parts. In the first part, using co-integration techniques, we analyze the long-term co-movements of prices at different hubs of the natural gas network. In the long-run in a well-integrated network, the price difference between any two nodes should be due to transportation cost differences, not to space supply concentration or other economic and environmental factors, if they are part of the same market. These other factors are assumed to be stationary processes. At the same time, spot gas prices are best approximated by processes containing a unit root (see the Appendix). In econometric terminology, if the difference between two non-stationary series is caused by stationary variables, the series are co-integrated. We attribute lack of co-integration of the prices at any two nodes to transmission problems, i.e., lack of technical means to ship more to one and less to the other node.

The second part of our analysis concerns the short-term dynamics, and is based on the impulse response technique.We analyze the way the different parts of the network respond to temporary local shocks. In a well-integrated network, or "market," a local price shock should spread across the network and disappear because of arbitrage. However, in the case of shocks at isolated nodes, they may take a very long time to affect other parts of the network, due to technical limits on shifting gas. Thus, by comparing timing of the responses of the different parts of the network to local shocks, one may obtain indications of segmentation of the network or the market.

There is a second, longer run version of the bottleneck, when adjoining receiving nodes do not carry out transactions in pipeline transit space that are alternatives to each other. In this example, gas to New York is never shifted to Chicago, because the Henry Hub has no interconnection between New York and Chicago. Then these two receiving nodes are not in the same market.

Our results show that the national collection of networks of different pipeline companies comprises three separate markets: Northeastern, with central receiving hubs in the New York City area, Midwest with its center in Chicago and in a Western market, primarily California. Section Two describes the methodology and data used. Section Three reports the result of co-integration and impulse response analysis, Section Four provides the conclusion. The results of the unit root tests for the spot prices are reported in the Appendix.

2.2. Methodology and Data Description

2.2.1. *Methodology*

The analysis is divided into two parts. The first part focuses on long-run co-movements of prices at the different nodes. This part requires the unit root[3] type of non-stationarity of the price series, which allows us to utilize the co-integration technique for detection of bottlenecks in the network. The results of unit root testing reported in the Appendix support the unit root hypothesis for the spot prices.

---

[3]A time series with unit root is described by the following equation: $x_t = x_{t-1} + \varepsilon_t = \sum_{s=0}^{\infty} \varepsilon_{t-s}$, where ε_t is a stationary shock in period t. In other words, at any given point of time, the value of a time series is the result of accumulation of shocks over its entire history.

We assume that each local hub is driven by unobservable non-stationary factors that determine prices in the long run. Further, we assume that, in the long-run equilibrium, a well-integrated network or "market" should be characterized by a single non-stationary pricing factor for all its hubs, and the difference between any two price series should be explained by local factors stationary by nature, such as weather/seasonal factors, or random local demand shocks. If there were non-stationary factors responsible for price differences, prices at different hubs would have deviated from each other for prolonged periods of time, and generated arbitrage opportunities. In other words, in the long-run, prices at different parts of a well-integrated network have to demonstrate similar dynamic patterns to be in the same market.

We assume that price at hub k in period t can be decomposed as:

$$p_{kt} = \theta_{0k} + \theta_{1k} f_{kt} + \ell_{kt}, \tag{2.1}$$

where p_{kt} is the real price at hub k, f_{kt} is a non-stationary long run pricing factor, and ℓ_{kt} is a local, stationary factor. For a network to be well integrated, f_{kt} has to be the same for all hubs: $f_{kt} = f_t$ for all k. In such a case, for every two price series, there exists a linear combination that eliminates the common factor f_t. Such series are called co-integrated. Thus, co-integration implies the absence of arbitrage opportunities in the long-run.

There are numerous co-integration tests. One that suits our purpose is the test based on the procedure developed by Kwiatkowski *et al.* (1992, KPSS). The procedure was originally developed for unit root testing, but it can be extended to co-integration as well (Tanaka, 1996). The KPSS co-integration test assumes that under the null hypothesis two time series are co-integrated. Thus, contrary to many other procedures, we would reject the hypothesis that two price series are co-integrated, and the network is well integrated, only if there is strong evidence in support of bottlenecks.

Our analysis is performed as follows: we regress the real prices at a "destination" hub on the real prices at a "source" hub. If the source and destination have different pricing factors then the residuals would contain a stochastic trend, which can be detected by the KPSS stationarity tests. Large values of the KPSS test statistic (or small values of the corresponding P-values) indicate the presence of a unit root in the residuals, i.e., a bottleneck on the "spoke" that connects source and destination hubs.

The slopes and intercepts from these regressions provide useful information as well. The intercept is an estimator of the transportation price for service between source and destination hubs. The slope coefficient evaluates conformity to the Law Of One Price, given that any increase in the source price of gas should result in the same increase in the destination spot price of gas for sale. Thus, if the slope is equal to one the Law is confirmed in a particular direction. But slope values greater than one then indicate a shortage of transmission capacity to be identified with segmentation, and values less than one indicate that the destination hub has access to other cheaper sources of natural gas, and thus the effect of any price increase at this source hub is partially smoothed by switching to other sources.

The second part of our analysis considers the short-run relations between prices across the network. We utilize the vector-auto-regression (VAR) and impulse response methodology to observe transmission of shocks through the natural gas network. In particular, the impulse response technique enables us to observe the effect of a shock in a price series at hub i on a price series at hub j after r periods. For example, zero values of impulse responses for $i \neq j$ would indicate that the shocks are not spread across the network, and the hub markets are isolated from each other.

This analysis in the second part is performed on basis differentials, which are constructed as differences between the gas price at the end point of a pipeline and the gas price at the origin in some production region. Let $d_t = (d_{1t}, \dots d_{kt})'$ be a vector of k basis differentials at time t, and $s_t = (s_{1t}, \dots s_{kt})'$ be the k-vector of associated shocks.

It is assumed that every basis differential depends on past and present values of all k shocks: $d_t = B(L)s_t$,where $B(L)$ is a polynomial in the lag operator i.e., $B(L) = \sum_{i=1}^{\infty} \beta_j L^j$ where $Ls_t = s_{t-1}$. Impulse responses are given by $\partial d_{i(t+r)}/\partial s_{jt}$, which is exactly the effect of a shock to series i on series j after r periods. In practice, impulse responses are calculated by, first, estimating a finite order VAR model $C(L)d_t = s_t$, and, then, inverting it to the moving average form $d_t = B(L)s_t$.

2.2.2. *Data description*

The database has been extracted from Natural Gas Intelligence (Intelligence Press Inc., Sterling Virginia) that has daily, monthly (average of the bidding week prices) and annual natural gas spot price data for 85 major hubs in the US from January, 1988 to December, 2003. Prices in the database are nominal; we estimate real prices by adjusting them by the index of prices of industrial production (*source*: IFS, IMF). We use the monthly data, because the majority of pipeline space contracts are made during "bid week", and, in the rest of the month, the volume of trade in spot space is small, and usually agents trade only for adjustments to the contracts already made. In addition, we assume that gas prices during bid week reflect the "real" situation in the market, i.e., trades are in the "thick" not "thin" period of exchange at source and destination hubs.

2.3. Results

2.3.1. *Co-integration*

In Table 2.1, we report the KPSS test statistics and their P-values. First, consider the line from Canada to California. Gas flows from Kingsgate to Stanfield and from Stanfield to Malin, which is on the border of California. For gas flows between Kingsgate and Stanfield, the KPSS test statistic is small, and the slope coefficient is close to one, though significantly less than one. However, as we approach the Californian border, we observe a distinct bottleneck between Stanfield and Malin with a large KPSS statistic and the slope coefficient being greater than one, which indicates insufficient transmission capacity. The flow of gas inside California, between Malin and PG&E City gate, encounters another bottleneck with high KPSS and slope greater than one.

The second major source for gas deliveries to California originates in West Texas, with a bottleneck between Waha and the Southern California border at Topock demonstrating a high KPSS statistic and slope significantly greater than one. Together Stanfield and Waha indicate that a "Californian market" is separated from the rest of the national gas

Table 2.1. Bottleneck Analysis of the Regional Natural Gas Transmission Network

Line	Source	Destination	KPSS Test Statistic	P-value	CI of Slope		CI of Intercept	
Rocky Mountains	Kingsgate	Stanfield	0.2658	0.0660	0.9248	0.9674	0.1031	0.2220
Rocky Mount. — Calif.	**Stanfield**	**Malin**	**14.2088**	**0.0000**	**1.2465**	**1.4282**	**−0.8197**	**−0.2426**
California	**Malin**	**PG&E CG**	**3.5414**	**0.0000**	**1.1673**	**1.2915**	**−0.6261**	**−0.0992**
California	PG&E CG	Kern River, PG&E	0.2216	0.1003	0.9843	1.0955	−0.4010	−0.1031
West TX – Calif.	**Waha**	**South. CA Border**	**1.8177**	**0.0000**	**1.0275**	**2.8638**	**−6.0135**	**0.634**
East TX – LA	Katy	Henry Hub	0.2845	0.0593	0.9372	1.0065	0.0574	0.2598
East TX – Midwest	Katy	Chicago CG	0.6710	0.0033	0.9367	1.0264	0.1054	0.3673
West TX - Midcont-t	El Paso Perm.	NNM 10-13 (KS)	0.0899	0.4713	0.9269	0.9673	0.0072	0.1530
Midcon — Midwest	**NNM 10-13 (KS)**	**Chicago CG**	**0.582**	**0.0063**	**1.0406**	**1.0955**	**−0.0131**	**0.1793**
Midcon — Midwest	**ANR SW**	**Chicago CG**	**1.7035**	**0.0000**	**0.9034**	**0.9915**	**0.3227**	**0.5507**
LA – Midwest	**Henry Hub**	**Chicago CG**	**1.8794**	**0.0000**	**0.9738**	**1.0177**	**0.066**	**0.1894**
LA – Midwest	**ANR SE, LA**	**Detroit**	**0.7739**	**0.0017**	**0.9832**	**1.0513**	**0.0977**	**0.2947**
LA – Northeast	Henry Hub	Alabama	0.1086	0.3573	1.0681	1.2487	−0.3956	0.2636
LA –Northeast	Alabama	NY	0.1207	0.3030	1.0302	1.3483	−0.5628	0.5978
Midwest – NE	Chicago	NY	0.1806	0.1473	0.9908	1.2782	−0.3963	0.6751

Large values of KPSS statistics and correspondingly low P-values indicate the presence of bottlenecks. The lines with P-values less than five percent have bottlenecks and are highlighted in bold. CI of slope/intercept provides Confidence Intervals for both Slope/Intercept. We abbreviate Northern Natural Mid 10-13 (Kansas) hub as NNM 10-13 (KS) and midcontinental hub as "midcon".

transmission network. There is insufficient transmission capacity on the lines that lead to California. The bottlenecks have been detected at the borders, which can be partially explained by the breakdown of one of the El Paso pipelines in 2000–2001.

Next, we consider the region that consists of spokes from Texas and Louisiana hubs to Midwest hubs close to major consuming centers, such as Chicago and Detroit. Between El Paso and Kansas the slope coefficient is significantly smaller than one, which indicates that other gas sources affect prices in the receiving hub. There is a bottleneck between Kansas and Chicago city gate. In this case, the estimated slope coefficient is close to one, the KPSS test statistic is relatively small, but significant. A similar pattern is observed for pipelines connecting the Louisiana producing hubs to the Midwest city gate hubs. That is, for originating hubs in Louisiana and receiving hubs in Chicago and Detroit, the KPSS statistics are much smaller than for California and slope statistics are close to one[4].

The major receiving hubs in the Midwest have a second major source of gas in Western Canada. In recent years, the importance of the Canadian source has been increasing, in

[4]The only serious bottleneck here is between Henry Hub and Chicago, however, it may be caused by the fact that gas from Henry Hub goes mostly to New York.

particular, because, in Canada, pipeline expansions and new construction have not been affected by price caps on space services in lines and, consequently, available capacity has been more responsive to the two recent gas price increases in the upper Midwest.

The third area of interconnection is pipeline capacity from Louisiana origin hubs to Northeast consuming hubs. No bottlenecks have been detected. However, on the line between the Alabama intermediate hub and New York receiving hubs, the slope coefficients tend to be significantly larger than one. This indicates greater price volatility in the destination than the origin, and the existence of friction in arbitrage of short term price spikes at specific locations.

Finally, we have investigated the possibility of substantial connection between the Northeastern and Midwest receiving nodes, sufficient to put them in the same "market." The last line of Table 2.1 shows that there are no bottlenecks between New York and Chicago. This result is somewhat surprising, since, contrary to New York, Chicago has access to substantial Canadian gas. In addition, according to our analysis, the Midwest has less connection with southern sources as compared to New York. Apparently, the originating hubs manage to arbitrage the price difference between New York and Chicago, which leads to same price dynamics at these two points in the long-run.

2.3.2. *Short-run (impulse response) analysis*

In this section, the short-term analysis is based on three basis differential series that represent three hypothetical Northeast, Midwest and Western (California) markets. A representative price differential for each of the three geographical areas was selected, as primarily determined by completeness of the data. Unfortunately, hub-to-hub price series have a large number of missing values, so that only the longest series with the least number of missing observations for each market could be used. We constructed basis differentials between Henry Hub and Columbia Gas (the largest retail distribution in the Northeast) El Paso to Chicago Citygate for the Midwest and Permian Basin to Southern California Border for the West.

In the first step, we estimated the VAR model for three series of basis differentials: North East (NEAST), Midwest (MWEST) and West (WEST). In the second step, the estimated VAR was inverted, and nine impulse response functions and their standard errors were computed. For the purpose of VAR estimation, we set the number of lags equal to five (however, the pattern of the impulse responses was approximately the same for different numbers of lags).

The main results are reported in Figure 2.1. The graphs in first column of Figure 2.1 show responses to the shocks originated from the Northeast; the second column shows

Figure 2.1. Shock Responses of Markets in Northeast, Midwest and West

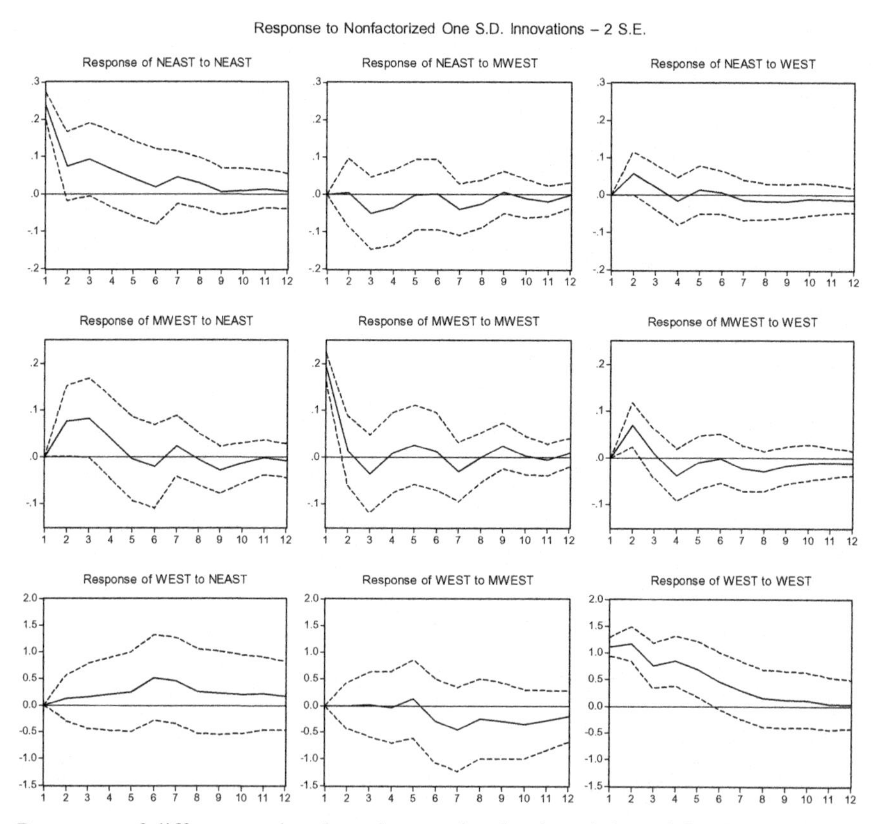

Response of different regional markets to the shocks originated from a particular local market plus-minus 2 standard errors. The first column shows response to shocks originating from Northeast, the second column to shocks from the Midwest, and the third column to shocks from the West. For Northeast and Midwest, the local shocks remain localized and do not affect other regions. The shock in the West persists for approximately half a year and significantly increases prices in the Northeast and Midwest for 1–2 months.

responses to the shocks originated from the Midwest, and the third column shows responses to the shocks originated from the West. It appears that the local shocks remain localized and do not have significant effects on spot gas exit prices in the other regions, with exception only for the shocks that originated in the West. A shock at the Northeast leads to a limited increase of prices in the Midwest, and almost no reaction in the West; the shock itself persists for approximately two months. Midwest shocks have no effect on either the Northeast or the West, and the shock also dies out after two months. The situation with the shocks originating from the West is completely different. Such a shock persists for approximately half a year and leads to significant increases in prices in both Northeast and Midwest for at least one to two months.

We conclude that, in the short-run, for periods of a few months or a heating season the national gas network is separated into three regional markets. Recall that the long-run analysis shows that the Midwest and the Northeast regions were integrated. While their prices tended to move together in the long-run, they did exhibit different patterns in the short-run with Midwest shocks having no effect in the East. This result is not particularly surprising, given the fact that the Midwest extracts gas from the pipelines coming from West Texas and Canada, whereas the Northeast takes gas from the hubs in East Texas and Louisiana. These two have contiguous but not identical sources and, thus, it takes longer for price differences to adjust.

2.4. Conclusion

Co-integration and impulse response techniques have been used to analyze the price behavior of natural gas across the network to determine whether local transactions in his space are integrated into common markets for gas transport and supply. We conclude that the natural gas network consists of three relatively isolated markets: the Northeast, Midwest, and California. While very large demand shocks originating in California have affected the rest of the country in the same heating season, there is no buy–sell mechanism that provides sufficient price adjustment across the network to make the country one market. As a result, in the long-run, the natural gas sales prices in California exhibit patterns different from those in the rest of the Country. And currently price adjustment between the Northeast and Midwest is rather limited. We attribute this segmentation to insufficient transmission capacity in the pipelines.

2.5. Appendix: Unit Root Test

In this section, we present the results of unit-root tests. We performed a series of Augmented Dickey–Fuller tests with the lag length equal to five. Table 2.2 summarizes the results. The null hypothesis of unit-root cannot be rejected at 5 percent significance for the vast majority of the price series. As we can see from Table 2.2, the smallest P-value is about 3 percent for only three series out of sixteen that

Table 2.2. Augmented Dickey–Fuller Unit Root Tests

Hub	**Augmented Dickey–Fuller Test Statistic**	**Approximate *P*-value**
Alabama	−2.8460	0.0521
Kern PG&E	−1.2530	0.6499
Ca Malin	−3.0620	0.0295
CaPGE gate	−3.1180	0.0253
ETX Katy	−2.0970	0.2457
ANR SW	−2.3800	0.1473
Chicago	−2.5230	0.1100
Detroit	−2.4160	0.1374
Kansas	−3.0230	0.0328
NE NY	−2.3160	0.1670
Kingsgate	0.6300	0.9866
Stanfield	−2.5520	0.1033
ANR SE, LA	−2.2550	0.1870
Henry Hub	−2.2550	0.1869
WTX	−2.2760	0.1798

we used for the analysis: CA Malin, CAPGE gate, Kansas. Thus, we maintain the assumption of unit root in our analysis.

References

Kwiatkowski, D., Phillips, P.C.B., Schmidt, P. and Shin, Y. (1992). "Testing the Null Hypothesis of Stationarity Against the Alternative of a Unit Root: How Sure Are We that Economic Time Series Have a Unit Root?" *Journal of Econometrics*, 54, 159–178.

MacAvoy, P.W. (2000). *The Natural Gas Market*, Yale University Press, New Haven.

Tanaka, K. (1996). *Time Series Analysis: Non-Stationary and Noninvertible Distribution Theory*, Wiley, New York.

CHAPTER 3

Quantitative Study Number Two

COMPETITION AMONG THE FEW IN THE NATURAL GAS PIPELINE INDUSTRY AFTER PARTIAL DEREGULATION

*Nickolay Moshkin**

This chapter considers the impact of regulatory changes during the 1990s on the competitiveness of pricing in gas transportation markets, measured in the changes that took place in contracts for gas and pipeline service. Of central concern are the effects on wellhead (supply) and

*Nickolay Moshkin is a Senior Manager at Cornerstone Research, New York. The material discussed herein may not reflect the opinions of Cornerstone Research.

city-gate (demand) wholesale gas prices, gas deliveries, and transportation charges. These determine payments by consumers for the consumption of natural gas.

The resulting changes are measured in prices and deliveries of the large three to six pipelines serving any of three wholesale distribution markets. To assess these changes, we introduce and estimate an integrated demand/supply model of the natural gas industry that explicitly accounts for the existing pipeline network architecture. The advantage of this model, in comparison with other economic models analyzing wellhead gas supply and city-gate gas deliveries, is that it takes into account geographic separation of supply at the wellhead and demand at the city gate. The model accounts for interactions between network entry and exit node transactions linked by a pipeline, in contrast to no interactions for unlinked states.

To illustrate, consider a simple example. ANR pipeline delivers gas from Texas to Illinois, Indiana, Iowa, Michigan, Ohio, and Wisconsin. A positive demand shock in Illinois would tend to increase wellhead prices in Texas producing areas from which ANR and other pipelines pump gas. That, in turn, would affect delivery prices in the above-listed states to which this pipeline or other pipelines (to which ANR is interconnected) deliver gas. The effect on other, unconnected, states or pipelines (for instance,

California and El Paso pipeline) is of a much lesser magnitude.

To our knowledge, the structure of the network has not been explicitly examined in previous research. Performance as well as structural tests can be undertaken. But accounting for network architecture and natural gas flows provides partial response to questions about the competitiveness of pipeline service. Open access to gas transportation, part of the deregulatory process, induced entry into the transportation sector of the market by brokers and dealers, and that entry significantly affected prices.

Using traditional econometric tests, we determine that transportation charges in the partial deregulatory period declined by approximately 12 percent compared to the regulatory period. Total annual savings to consumers from the new open-access orders when estimated by using the price change times consumption quantity amounted to $1.93 billion per year. The structural model provides a lower estimate of about $1.05 billion per year.

The model is also used to analyze the impact of a new pipeline on incumbent pipelines' prices and quantities. As an example, we choose the newly proposed pipeline from the Alaska North Slope, and, after calibrating the necessary pipeline-specific variables at the values found elsewhere in the press or in government publications, we simulate the resulting competitive effect on the prices and quantities that such a pipeline would introduce. We find that the construction of such a pipeline is economically appealing, as it would generate an estimated total of $0.78–1.62 billion in annual consumers' savings as a result of the price decline at the city-gate.

The remainder of the chapter is organized as follows. Section One describes the data. Section Two provides the econometric specification for the various traditional tests of structural change. Implementation details and results are reported there as well. Section Three lays out the integrated demand/supply model of the natural gas industry. Section Four presents model estimation results and discusses their implications and validity. Section Five presents two policy evaluations based on model estimates from our models: the first estimates the economic benefits of the Orders 636 *et al.*, and the second forecasts the economic effects that the construction of the Alaska pipeline will have on the market. Section Six concludes.

3.1. Data Description

The data set covers the 1990–1997 period, with monthly or annual observations for the variables. The data includes monthly prices, quantities, and other characteristics of the natural gas market. The original, non-transformed, data contain the following series:

1. The average monthly prices of natural gas at the origin node, by pipeline.[1]
2. Annual interstate deliveries, by pipeline, 1990 and 1994–1997.[2]
3. The average monthly prices of natural gas by state at the exit node where the gas is transferred from a pipeline to a local distribution company within the state (i.e., "city-gate" price).[3]
4. Monthly natural gas deliveries to all consumers, by state.[4]

---

[1]*Source*: Federal Energy Regulatory Commission's publication "*Inside the FERC*," various months.

[2]*Source*: Energy Information Administration, Form 2, various years.

[3]*Source*: Taken originally from forms EIA-857, "*Monthly Report of Natural Gas Purchases and Deliveries to Consumers*," and reported in various issues of EIA publication "*Natural Gas Monthly*."

[4]*Source*: Taken originally from forms EIA-857, "*Monthly Report of Natural Gas Purchases and Deliveries to Consumers*," and forms EIA-759, "*Monthly Power Plant Report*," and reported in various issues of EIA publication "*Natural Gas Monthly*."

5. Monthly marketed production of natural gas, by state.[5]
6. Average monthly temperature, by state.[6]
7. State population estimates: annual time series, July 1, 1990 to July 1, 1997.[7]
8. Average annual wage, by state.[8]
9. Annual natural gas reserves, by state.[9]
10. Distances measured "as the crow flies" between the largest cities of each state.[10]

To estimate the supply side of the model, we focus on the average monthly wellhead prices for 21 major interstate pipelines. This set contains virtually all major interstate

---

[5]*Source*: Taken originally from form EIA-895, "*Monthly Quantity of Natural Gas Report*," EIA calculations, and reported in various issues of EIA publication "*Natural Gas Monthly*".

[6]*Source*: 1990–1994: Teigen, Lloyd D., Weather in US Agriculture (computer file). #92008B. Washington, D.C.: Economic Research Service, US Dept. of Agriculture, January 1992 (updated January 1996). 1995–1997: US Department of Commerce, National Climatic Data Center publications "*Historical Climatology Series*".

[7]*Source*: Population Estimates Program, Population Division, US Census Bureau, Washington, DC 20233.

[8]*Source*: Bureau of Economic Analysis web page, file http://www.bea.doc.gov/bea/regional/reis/ca34/index.htm.

[9]*Source*: EIA publication "*US Crude Oil, Natural Gas and Natural Gas Liquids Reserves 1998 Annual Report*".

[10]*Source*: Calculated using web page http://www.indo.com/distance/.

pipelines that operate in the United States market.[11] To estimate the demand side of the model, we collected monthly quantities demanded, by state, and the average monthly city-gate prices, by state, for all US states. However, we excluded 22 states from the analysis for the following reason: a) 10 producing states, in which marketed production exceeded deliveries to consumers; b) three states that had no interstate pipelines delivering gas; and c) nine states with small natural gas demand, with deliveries to consumers of less than 100 billion cubic feet in 1999. This results in a sample of 28 consuming states and 21 major pipelines.

Tables 3.1 and 3.2 report the descriptive statistics of the gas prices and quantities. The means and variances of the wellhead prices are presented in Table 3.1, and the means and variances of the quantities demanded and city-gate prices are presented in Table 3.2.

The supply side of the model is associated with individual pipelines, and not with producing states. Therefore,

---

[11]The only noticeable exception is Pacific Gas Transmission Co., which during this period delivered about 35 percent of natural gas to California. The pipeline was excluded due to the lack of data. We believe that the exclusion does not introduce a systematic bias into the model results.

Table 3.1. Monthly Wellhead Prices, by Pipeline

Pipeline	**Price, $/Mcf**			
	Mean	**Std**	**Max**	**Min**
ANR Pipeline Co.	1.79	0.48	3.55	1.01
CNG Transmission Corp.	2.32	0.66	4.55	1.39
Columbia Gas Transmission Corp.	2.13	0.58	4.20	1.28
Columbia Gulf Transmission Co.	1.83	0.50	3.50	1.00
El Paso Natural Gas Co.	1.59	0.49	3.63	0.95
Florida Gas Transmission Co.	1.88	0.49	3.60	1.02
Kern River Gas Transmission Co.	1.52	0.54	3.81	0.78
Koch Gateway Pipeline Co.	1.72	0.44	3.23	0.94
Natural Gas Pipeline Co. of America	1.75	0.46	3.56	0.99
NorAm Gas Transmission Co.	1.72	0.46	3.70	0.97
Northern Natural Gas Co.	1.68	0.48	3.74	1.00
Northwest Pipeline Corp.	1.39	0.49	3.74	0.74
Panhandle Eastern Pipeline Co.	1.70	0.46	3.68	1.00
Southern Natural Gas Co.	1.84	0.49	3.54	1.01
Tennessee Gas Pipeline Co.	1.79	0.47	3.44	1.00
Texas Eastern Transmission Corp.	1.81	0.47	3.48	1.02
Texas Gas Transmission Corp.	1.85	0.50	3.60	1.02
Transcontinental Gas Pipeline Corp.	1.84	0.49	3.49	1.02
Transwestern Pipeline Co.	1.68	0.45	3.69	1.03
Trunkline Gas Co.	1.70	0.40	2.90	0.98
Williams Natural Gas Co.	1.68	0.48	3.86	1.00

Table 3.2. Monthly Quantities Demanded and Prices, by State

State	**Quantity, MMcf**		**Price, $/Mcf**			
	Mean	**Std**	**Mean**	**Std**	**Max**	**Min**
Arizona (AZ)	8,570	2,074	2.49	0.43	3.78	1.56
Arkansas (AR)	2,530	3,778	2.46	0.30	3.75	1.78
California (CA)	127,910	22,380	2.58	0.41	3.71	1.72
Connecticut (CT)	9,693	3,721	4.12	0.64	5.54	2.79
Florida (FL)	32,900	7,683	2.82	0.50	4.32	2.16
Georgia (GA)	28,446	8,602	3.35	0.52	4.83	2.19
Illinois (IL)	84,654	2,933	2.96	0.42	3.96	2.15
Indiana (IN)	42,124	16,863	2.91	0.40	3.66	1.85
Iowa (IA)	19,700	8,147	3.28	0.87	7.22	2.30
Kentucky (KY)	8,715	7,263	3.01	0.37	3.96	2.23
Maryland (MD)	15,293	5,775	3.62	0.85	5.69	2.33
Massachusetts (MA)	26,579	8,136	3.70	0.69	5.61	2.54
Michigan (MI)	54,862	33,755	2.81	0.34	3.57	2.10
Minnesota (MN)	26,261	2,616	2.86	0.45	4.11	2.02
Mississippi (MS)	9,962	2,983	2.71	0.38	3.91	2.05
Missouri (MO)	21,508	12,071	3.22	0.60	4.85	2.15
Nebraska (NE)	10,068	4,424	3.12	0.82	6.41	2.03
Nevada (NV)	7,724	2,274	2.84	0.54	4.65	2.04
New Jersey (NJ)	44,682	16,807	3.37	0.43	4.58	2.58
New York (NY)	84,886	26,862	2.91	0.43	3.99	1.97
North Carolina (NC)	15,325	4,949	3.08	0.38	3.93	2.18

(*Continued*)

Table 3.2. (*Continued*)

State	Quantity, MMcf		Price, $/Mcf			
	Mean	**Std**	**Mean**	**Std**	**Max**	**Min**
Ohio (OH)	57,357	33,172	3.70	0.83	7.42	2.59
Pennsylvania (PA)	43,189	24,047	3.54	0.52	4.91	2.44
South Carolina (SC)	11,721	2,048	3.32	0.42	4.15	2.01
Tennessee (TN)	19,640	7,964	2.96	0.55	5.98	1.72
Virginia (VA)	14,099	6,143	3.20	0.55	4.67	2.06
Washington (WA)	16,306	5,386	2.14	0.36	3.24	1.63
Wisconsin (WI)	29,478	14,200	3.50	0.76	6.09	2.30

we transform variables that influence supply (for example, proved gas reserves) and the interaction between supply and demand (for example, distance between the fields) from being defined at the state level to being defined at the pipeline level. In order to accomplish this, we make an assumption that the long term production-to-reserves ratio is constant across pipelines in each state. By using historical gas transportation volumes, we effectively reassign state reserves to pipelines in proportion to pipelines' gas volumes from the state. Similarly, we define the length of a pipeline to a particular consuming state by weighting distances between producing states with weights proportional to volumes.

In such a transformation of variables, we create a matrix of interstate transportation quantities, by pipeline and by state. We call this matrix *the transportation matrix.* The transportation matrix allows us to reverse-engineer otherwise unavailable variables (e.g., monthly transportation quantities to each state by each pipeline) and to generate instrumental variables required for identification of the model.

For a standard simultaneous equation model of supply and demand, that does not have a feature of geographical separation of supply and demand, the variables to identify the model are defined and applied in the typical fashion: the variables included in the demand equation but not in the supply equation (pure demand shifters) are used to identify the supply equation, and vise versa. However, given that our model takes into account the geographical separation of supply and demand, it is not immediately obvious how to define and use these variables. For instance, should an average temperature for the entire United States or a temperature at a particular location of the pipeline network be used as a demand shifter to identify the supply relationship for a pipeline? It seems natural to use some *weighted average* of the values of the variable of interest at various network locations that are relevant to a pipeline to specify such a variable. Here, we use the weights derived from the transportation matrix.

To illustrate how this works, let a pipeline deliver 70 percent of its gas to state A, 30 percent to state B, and does not deliver any to state C. Let the value of a demand shifter X^D at the three states be given by a vector:

$$X^D = \begin{bmatrix} X_A^D \\ X_B^D \\ X_C^D \end{bmatrix}. \tag{3.1}$$

Then the new "network transformed" (i.e., *projected*) variable X_{Pipeline}^D, which can be used to identify the supply equation of the pipeline, is defined as 70 percent of X_A^D plus 30 percent of X_B^D. Note that due to the absence of the pipeline in state C, the value of the variable there has no effect on the value of the variable for the pipeline. Similarly, for supply shifters by state, we use pipelines' shares in these states as weights for averaging.

The model does not allow for gas storage. We assume that all gas that is consumed in a particular month by a particular state that is not produced within the state has to be transported during that month by pipeline companies operating in the state. Therefore, the quantities delivered to all consumers less marketed production by state, where available, provide the quantities transported to states. Using these and annual pipeline shares in each state, we derive the quantities transported monthly by each pipeline.[12]

3.2. Traditional Econometric Tests

3.2.1. *Reduced form regressions*

To assess the effects of partial deregulation, we divide the data into two equal time periods: January 1990 to December 1993, and January 1994 to December 1997. The split is appropriate for the purpose of finding the effects of the Orders 636 *et al.*, which divested gas ownership from pipeline operations, and price capped pipeline space only on primary contracts. The orders went into effect at the beginning of the 1993–1994 heating season. As a first set of tests, we run reduced form regressions of prices and quantities to find the economic significance of various explanatory variables. The explanatory variables for the demand side are monthly temperature, population, per capita income, price of crude oil, projected gas reserves, population and per capita income in the states competing for the same gas reserves as the state in question, and a binary variable for

[12]For a more detailed explanation on how these transformations of variables defined at the demand locations and of those variables defined at the supply locations (and vice versa) were implemented please see the Appendix.

deregulation (equal to 0 for the regulatory period, and 1 for the partial deregulatory period). The explanatory variables for the supply side are gas reserves, price of crude oil, projected monthly temperature, population and income, and the same binary variable for deregulation.

Table 3.3 reports the results of the linear regression for the demand side variables.[13] Most if not all coefficients in the first regression of quantities demanded by states are of expected sign. *Ceteris paribus*, demand is higher in the winter when the temperature is lower; higher populated states demand more; wealthier states demand more; states that have more projected reserves demand more; states that have higher populated or wealthier states competing for the same reserves demand less; and, finally, demand declines in response to an increase in the price of crude oil. The average quantity demanded in consuming states increased in the partial deregulatory period by about 4.7 billion cubic feet per month per state, which constitutes about a 10.2 percent increase. All coefficients are statistically significant. This finding of the negative relationship between the price of crude oil and the quantity demanded of natural gas

---

[13]We have run quadratic regressions that included linear terms, cross products and second order terms of all variables. The results are quite similar and thus not reported here.

Table 3.3. Reduced Form Regressions, Demand Side

Variable	**Quantity Demanded, Bcf**		**City-Gate Price, $/MMcf**	
	Coefficient	**t-statistic**	**Coefficient**	**t-statistic**
Constant	48.174	12.17	10.72	0.06
Temperature, °C	−0.729	−42.88	7.62	10.03
Population, Mil	4.417	69.02	−12.60	−4.48
Income, $1000	0.563	4.65	31.65	5.93
Projected Reserves, Tcf	1.314	12.40	27.26	5.82
Population, others, Mil	−0.541	−7.03	−41.01	−12.03
Income, other, $1000	−1.043	−4.78	86.53	9.00
Price of Oil, $/barrel	−0.290	−3.12	1.50	0.37
$\text{Dummy}_{1994\text{–}1997}$	4.655	5.76	−286.71	−8.04
Number of Observations	2,660		2,660	
R^2, %	80.5		18.0	

calls for a special consideration of the interaction between the oil and gas markets. At first, it seems natural to treat the price of crude oil as a demand shifter and include it in the demand equation assuming that oil and gas are substitutes in consumption. If the assumption were true, the coefficient on the price of oil would come out positive in the reduced form regressions rather than negative.

The contradiction might be resolved by taking a closer look on the industry infrastructure. Twenty-five percent of natural gas production comes from the oil wells. Therefore, the change in the price of the crude oil influences the firm-specific decisions of a company that operates oil-gas wells and extracts both products. However, the sign of the effect can go in both directions. On one hand, the increase in the price of crude oil might induce companies to intensify exploration activity, find more oil-gas wells, and increase production of both products. On the other hand, such price increase might make the oil extraction from the existing wells more economically attractive for the company and induce a shift of the company's limited capital and labor resources from gas to oil production. The reduced form regressions support the second hypothesis; and the effect dominates the small (if any) substitution effect in consumption in the demand equation. Consequently, in the integrated demand/supply model we include the price of crude oil variable in the supply equation, but not in the demand.

In the second regression of the city-gate price, some estimates are of the opposite sign from what one would expect. *Ceteris paribus*, price is higher in the summer, in lower populated states, in states with higher per capita annual income, with more projected reserves, or in lower populated or wealthier states competing for the same reserves, or when the price of crude oil is higher. The sign of the coefficient on the price of crude oil is also consistent with the multi-output firm story. The average city-gate price declined in the partial deregulatory period by about 28.7 cents per thousand cubic feet, or by about 9.6 percent. All coefficients but the one on the price of crude oil are statistically significant.

Table 3.4 shows the reduced form regression results for the supply side variables. Here, in the equation for the quantity supplied the sign of some coefficients are as one would expect, but others are not. More reserves imply more supply; lower temperatures, lower population and higher wages in the consuming states connected to a pipeline encourage higher supply; lower prices of crude oil increase gas supply. According to these estimates, the average quantity supplied by pipelines decreased in the partial deregulatory period by 5.4 billion cubic feet per month per pipeline, constituting a 12.8 percent decline. This stands in contrast to the findings from the reduced form regression of the quantities demanded — after partial deregulation, demands increased while supplies decreased. It would seem that further, more complex analysis integrating the two sides of the market is necessary to resolve the inconsistency.

Table 3.4. Reduced Form Regressions, Supply Side

Variable	Quantity Supplied, Bcf		Wellhead Price, $/MMcf	
	Coefficient	**t-statistic**	**Coefficient**	**t-statistic**
Constant	6.456	1.38	840.7	6.47
Reserves, Tcf	3.875	41.67	−10.82	−4.18
Projected Temperature, °C	−0.879	−35.16	−7.865	−11.42
Projected Population, Mil	−0.279	−4.10	−10.76	−5.72
Projected Income, $1000	2.474	14.47	45.88	9.66
Price of Oil, $/barrel	−0.317	−2.42	21.48	5.88
$\text{Dummy}_{1994\text{–}1997}$	−5.432	−5.40	41.30	1.47
Number of Observations	1,907		1,907	
R^2, %	68.5		16.3	

The results of the price regression are more intuitive. Gas field supply prices are higher during the colder time of the year, for pipelines that have smaller amounts of reserves, that deliver to the less populated or wealthier states, and when the price of crude oil is higher. In the partial deregulatory period the average wellhead price increased by 4.1 cents per thousand cubic feet, or by about 2.3 percent.

All coefficients but the one on the binary variable are statistically significant.

To sum up the preliminary reduced-form regression results, two out of four regressions, wholesale quantity market demanded and wellhead prices, have signs on the explanatory variables consistent with our intuition. The results show that consumption in the partial deregulatory period increased, city-gate price decreased while wellhead prices increased, presumably as a result of decreasing transportation industry markups. The plausible explanation for these findings is that restructuring increased competition in the transportation market, which, in turn, caused a movement of the prices and quantities down the demand curve. Further analysis is warranted to resolve some revealed inconsistencies.

3.2.2. *Basis differentials in prices*

The implicit transportation charge can be estimated by the difference between the gas price at destination, or city gate, and gas price at origin, or wellhead. This difference is called *the basis differential*. The Law of One Price with arbitrage opportunities within a market for a homogeneous product ties the prices at different points of a network market together. The Law asserts that if the product is regularly delivered from location A to location B, the difference between prices, $P_B - P_A$, should equal the service charge for moving the product from A to B. If it were not, and the difference were lower than the transportation charges, no firm would be willing to ship the product and incur losses, and diminishing supply of the product at location B would drive $P_B - P_A$ up to the level of the transportation charge. On the other hand, if the differential were higher than the transportation charge, market participants would arbitrage $P_B - P_A$ down to the transportation charge by increasing deliveries to node B. In any case, one does not expect to observe a systematic violation of the law, so that this relationship between gas price differences and transportation charges can be utilized to estimate changes in the latter caused by deregulatory Orders 636 *et al.*

For this estimation process, we generate two alternative sets of data. The first set includes differences between city-gate and wellhead gas spot prices for the physically connected pipeline-state pairs only. The second set includes all possible pipeline-state pairs. We run a regression of this basis differential on a set of supply and demand variables, length and length squared of the segment of the pipeline that delivers gas to the state, and the partial deregulatory period binary variable.

Table 3.5 reports the results. There are some differences between the estimates from the two data sets.

Table 3.5. Reduced Form Regressions, Basis Differential

Variable	**Basis Differential, $/MMcf**			
	Actual network pairs		**All possible pairs**	
	Coefficient	**t-statistic**	**Coefficient**	**t-statistic**
Constant	288.6	3.42	202.7	5.18
Temperature, °C	3.670	3.12	11.929	39.76
Population, Mil	−16.301	−11.87	−28.561	−53.49
Income, $1000	−6.831	−2.57	46.659	47.86
Reserves, Tcf	−4.029	−2.34	1.528	2.20
Projected Temperature, °C	11.209	9.56	3.819	12.16
Projected Population, Mil	−9.150	−6.15	6.639	12.97
Projected Income, $1000	20.296	6.27	−20.203	−15.34
Price of Oil, $/barrel	−17.484	−9.10	−18.255	−19.38
Distance, miles, 1000	655.2	7.88	982.6	35.47
Distance^2	107.2	1.92	−389.6	−33.02
$\text{Dummy}_{1994–1997}$	−171.6	−10.73	−179.3	−23.59
Number of Observations	9,614		53,480	
R^2, %	24.6		20.2	

The majority of the coefficients in both regressions have the expected sign. The results also reveal that transportation charges have quantity discounts, possibly due to the presence of increasing returns to scale in gas transportation. Implied (basis differential) charges are lower when the temperature is lower, namely, when the quantities transported are higher. Similarly, the charges are lower for more populated states. Wealthier states consume more gas, and pay less for gas transportation. Higher gas reserves imply higher quantities transported, and, in line with the former conjecture, lower transportation charges. However, the opposite relation is found with the data set that consists of all possible pairs, so network architecture matters.

The price differentials increase with distance between the origin and destination markets, in a convex fashion for the actual connections set, and a concave fashion for the all possible set. All but one coefficient are statistically significant. Both regressions show a statistically significant decline in transportation charges in the partial deregulatory period of 17 to 18 cents per thousand cubic feet of gas transported, which constitutes a decline of nearly 12 percent.

3.3. An Integrated Demand/Supply Model

These preliminary estimation results encourage further development of a network model. We introduce such a model that by incorporating pipeline network architecture, integrates geographically separated demand and supply markets and generates a system of simultaneous equations to explain transport performance. The model identifies the appropriate markups in the transportation segment of the industry and analyzes the effects that partial restructuring in the 1993 deregulatory process had on them. It also estimates the supply and demand elasticities and provides a framework for analysis of further regulatory policy change.

The system consists of three sets of equations and one set of identities. The first set classifies the demand for natural gas in each state. The second outlines the supply by each pipeline. The third provides the relation between the wellhead and city-gate prices. Finally, the identity ties quantities transported by pipelines with the quantities received by states, assuming that no losses occur within the network.

The first set of equations that determines each state's demand is given by an inverse demand function:

$$P_{jt}^{D} = \alpha^{D} + X_{jt}^{D}\beta^{D} - Q_{jt}^{D}\gamma^{D} + \varepsilon_{jt}^{D} \tag{3.2}$$

for $j \in \Omega_D$ and $t = 1, \ldots, T$. Here P_{jt}^{D} and Q_{jt}^{D} are city-gate price and quantity demanded by state j in period t; X_{jt}^{D} are demand shifters (temperature, population, and per capita income); ε_{jt}^{D} are idiosyncratic random shocks, independently and identically distributed (i.i.d.) across time and regions; and Ω_D is a set of consuming states.

The second equation that determines each pipeline's supply is given by an inverse supply function:

$$P_{it}^{S} = \alpha^{S} + X_{it}^{S}\beta^{S} + Q_{it}^{S}\gamma^{S} + \varepsilon_{it}^{S} \tag{3.3}$$

for $i \in \Omega_S$ and $t = 1, \ldots, T$. Here P_{it}^{S} and Q_{it}^{S} are wellhead price and quantity supplied by pipeline i in period t; X_{it}^{S} are supply shifters (gas reserves and price of crude oil); ε_{it}^{S} are idiosyncratic random shocks, i.i.d. across time and pipelines; and Ω_S is a set of pipelines.

The identity connecting quantities is:

$$Q_t^{D} = T_P' \cdot Q_t^{S} \tag{3.4}$$

for $t = 1, \ldots, T$. Here Q_t^{D} and Q_t^{S} are quantities demanded by states and transported by pipelines in period t, and T_P is a *pipeline normalized* transportation matrix for the same period.[14] This identity guarantees that all the gas

[14]The rows of the matrix contain ratios of a pipeline's deliveries to a particular state to total deliveries through that pipeline to all states. The Appendix describes in detail how the matrix was calculated.

injected in the network by pipelines has been taken out by the collection of all states.

To complete the system and to derive the remaining relationship between the wellhead and city-gate prices, we make the following assumptions on how the market is operating. First, we assume that a pipeline can choose the total amount of gas it transports in a given month, but not its relative allocation between the states to which the pipeline delivers. In other words, we assume that pipeline normalized transportation matrix T_P is *exogenously* determined by an engineering structure of the network. According to this assumption, when a pipeline adds one unit of gas at its entry point, its affiliates on the other end of the pipeline get the fractions of the unit in proportion to the corresponding row of T_P matrix.[15]

Second, we postulate that pipelines compete in accordance with a Cournot framework. Thus, each pipeline

---

[15]Indeed, this exogeneity assumption is restrictive and introduces some magnitude bias, although it does not change the direction of the pipeline reaction function. Due to this assumption, pipelines have limited flexibility to react on the individual demand shock in the consuming states, but have to adjust the total shipments to all recipients. Consequently, the pipeline's reaction is smaller than it would have been without this restriction.

chooses the transportation quantity to maximize its profits:

$$\max_{Q_{it}^S} \sum_{j \in \Omega_D} m_{ijt} Q_{it}^S \left(P_{jt}^D - ac_{ijt}(i, j, m_{ijt} Q_{it}^S, Q_{it}^S) - P_{it}^S \right) \tag{3.5}$$

for $i \in \Omega_S$ and $t = 1, \ldots, T$. Here m_{ijt} are the elements of the pipeline normalized transportation matrix T_P; Q_{it}^S is a total volume transported through pipeline i in period t; $m_{ijt} Q_{it}^S$ is the gas delivered to state j through pipeline i in period t; P_{jt}^D and P_{it}^S are the prices in state j and at the well-head of pipeline i; ac_{ijt} is an average cost of transporting one unit of gas to state j by pipeline i. Average costs might depend upon the pipeline and the state identities, quantity transported through the portion of the pipeline to the state, and the total volume transported by the pipeline.

We assume also that the average cost of transportation ac_{ijt} is independent of quantities transported, but is a quadratic function of the length of the link of the pipeline i to state j (plus an unobservable random component):

$$\begin{aligned} ac_{ijt} &= a_0 + a_1 L_{ij} + a_2 L_{ij}^2 + \varepsilon_{ijt} \\ ac_{ijt} &= c_{ij} + \varepsilon_{ijt} \end{aligned} \tag{3.6}$$

for $i \in \Omega_S$, $j \in \Omega_D$, and $t = 1, \ldots, T$. Here c_{ij} is a time-independent component of average transportation costs. We further assume that pipelines observe all realized random shocks before making corresponding transportation decisions. To achieve maximum profit, each pipeline chooses the transportation quantity that satisfies the following first order condition (FOC):

$$\sum_{j\in\Omega_D} m_{ijt}\left(P^D_{jt} - c_{ij} - \varepsilon_{ijt} - P^S_{it}\right) = \left(\gamma^D \sum_{j\in\Omega_D} m^2_{ijt} + \gamma^S\right) Q^S_{it} \qquad (3.7)$$

for $i \in \Omega_S$ and $t = 1, \ldots, T$. This equation asserts that the markup in Cournot oligopoly with constant average cost of transportation is proportional to the share of the total quantity transported in the market and is identified by the slopes of the inverse demand and supply functions. Therefore, the system consisting of the inverse demand equation (3.2), the inverse supply equation (3.3), the identity (3.4), and the pipeline's FOC for profit maximization (3.7) can be estimated together by Generalized Method of Moments (GMM), imposing cross-equation restrictions on some parameters.

Alternatively, let transportation exhibit increasing returns to scale, or the average cost be decreasing in the quantity transported in the following form:

$$\begin{aligned} ac_{ijt} &= a_0 + a_1 L_{ij} + a_2 L^2_{ij} - b_0 m_{ijt} Q^S_{it} - b_1 Q^S_{it} + \varepsilon_{ijt} \\ ac_{ijt} &= c_{ij} - b_0 m_{ijt} Q^S_{it} - b_1 Q^S_{it} + \varepsilon_{ijt} \end{aligned} \qquad (3.8)$$

for $i \in \Omega_S$, $j \in \Omega_D$, and $t = 1, \ldots, T$. This specification implies that average cost is diminishing both in quantity transported to the state j through the segment of pipeline i, $m_{ijt}Q^S_{it}$, and in total quantity transported by the pipeline to all states, Q^S_{it}.

In this case, (3.7) would change to:

$$\sum_{j\in\Omega_D} m_{ijt}\left(P^D_{jt} - c_{ij} - \varepsilon_{ijt} - P^S_{it}\right)$$
$$= \left((\gamma^D - 2b_0)\sum_{j\in\Omega_D} m^2_{ijt} + (\gamma^S - 2b_1)\right) Q^S_{it} \quad (3.9)$$

for $i \in \Omega_S$ and $t = 1, \ldots, T$. In contrast to the estimation of the above system of equations, no cross-equation restrictions can be imposed if (3.9) is included in place of (3.7). One can find coefficients of demand, supply and the new FOC (3.9) equations separately, and then deduce transportation costs coefficients b_0 and b_1 from the difference in the corresponding estimates.

3.3.1. *Conjectural variation*

One approach to measuring the state or extent of Cournot-type oligopoly directly through behavioral characteristics is to estimate the relevant conjectural variations (CV), the change in total rivals' output in response to a change in own output (see Iwata, 1974; Bresnahan, 1982; and Tirole, 1988). The conjectural variation measures the responsiveness of other firms to the firm's choices, and thus, represents a benchmark for the nature of competition in the market. We derive both (3.7) and (3.9) under the assumption that this parameter, $\lambda = \partial Q_{-i}^S / \partial Q_i^S$, is zero for all i. However, this should not necessarily be the case. The conjectural variation may be positive or negative. For perfect Cournot, firms do not take into account the effect of their actions on the other firms' choices, and, therefore, $\lambda = 0$. On the other hand, for Bertrand oligopoly, or perfect competition, a firm's quantity variation is offset by the response of other firms, total quantity remains constant, and, therefore, $\lambda = -1$. For collusion, all firms act in a perfectly correlated fashion, and hence the index is $\lambda = 1$. One way to measure the change in the level of competition or collusion within an industry is to estimate this parameter for sequential market time periods. MacAvoy (1995) estimates the CV parameter in the long-distance telecommunication market after the AT&T divestiture to assess whether this antitrust remedy had any effect on the competitiveness of the markets involved. He found that both MCI and Sprint implicitly collude ($\lambda > 0$), or price their services unaggressively, while AT&T behaves in accordance with Cournot ($\lambda \approx 0$).

Let us bring the notion of conjectural variation into the FOC (3.9):

$$\sum_{j\in\Omega_D} m_{ijt}(P^D_{jt} - c_{ij} - \varepsilon_{ijt} - P^S_{it})$$
$$= \left((\gamma^D - 2b_0) \sum_{j\in\Omega_D} m^2_{ijt} + \lambda\gamma^D \sum_{j\in\Omega_D} \left(m_{ijt} \sum_{k\neq i} m_{kjt} \right) + (\gamma^S - 2b_1) \right) Q^S_{it} \quad (3.10)$$

for $i \in \Omega_S$ and $t = 1, \ldots, T$. Here $\lambda = \partial Q^S_k / \partial Q^S_i$ for all $k \neq i$. In this case, when increasing returns to scale are present, and CV parameter is not constrained at zero, we are able to identify all the parameters of the system, but cannot impose any cross-equation restrictions.

3.3.2. *Structural system of simultaneous equations*

Based on the above specifications, we conduct the structural estimation of the system of simultaneous equations. The system consists of the inverse demand equation (3.2), the inverse supply equation (3.3), the identity (3.4) and the pipeline's FOC for profit maximization in the structural form (3.10). Let us denote $HHI^D_{it} = \sum_{j\in\Omega_D} m^2_{ijt}$ and $CHHI^D_{it} = \sum_{j\in\Omega_D} (m_{ijt} \sum_{k\neq i} m_{kjt})$. Equation (3.10) is written as:

$$\sum_{j \in \Omega_D} m_{ijt} \left(P_{jt}^D - c_{ij} - \varepsilon_{ijt} - P_{it}^S \right)$$
$$= (\gamma^S - 2b_1) \cdot Q_{it}^S + (\gamma^D - 2b_0) \cdot HHI_{it}^D Q_{it}^S$$
$$+ \lambda \gamma^D CHHI_{it}^D Q_{it}^S \quad (3.11)$$

for $i \in \Omega_S$ and $t = 1, \ldots, T$. To measure the effect of regulation on competition in the market, we allow the parameters of the equations to vary between periods. Thus, the net change in conjectural variation will provide an indication of the change in the competitive environment.

The system is given by the following set of equations and identity:

$$\begin{cases} \sum_{j \in \Omega_D} m_{ijt} P_{jt}^D = \alpha^D + \sum_{j \in \Omega_D} m_{ijt} X_{jt}^D \beta^D \\ \qquad + \sum_{j \in \Omega_D} m_{ijt} Q_{jt}^D \gamma^D \\ \qquad + \sum_{j \in \Omega_D} m_{ijt} \varepsilon_{jt}^D, \\ P_{it}^S = \alpha^S + X_{it}^S \beta^S + Q_{it}^S \gamma^S + \varepsilon_{it}^S, \\ \sum_{j \in \Omega_D} m_{ijt} (P_{jt}^D - a_0 - a_1 L_{ij} \\ \qquad - a_2 L_{ij}^2 - \varepsilon_{ijt} - P_{it}^S) \\ \quad = (c_0 + c_1 HHI_{it}^D + c_2 CHHI_{it}^D) \\ \qquad \cdot I\{Year \leq 1993\} \cdot Q_{it}^S \\ \qquad + (d_0 + d_1 HHI_{it}^D + d_2 CHHI_{it}^D) \\ \qquad \cdot I\{Year > 1993\} \cdot Q_{it}^S, \\ Q_t^D = T_P' \cdot Q_t^S. \end{cases} \quad (3.12)$$

3.3.3. *Estimation methodology*

We assume that $\varepsilon_{jt}^{D} \sim N(0, \sigma_D^2)$, $\varepsilon_{it}^{S} \sim N(0, \sigma_S^2)$, and $\varepsilon_{ijt} \sim N(0, \sigma^2)$. To estimate the system, we use a set of instruments Z_{it}^{D}, Z_{it}^{S}, and Z_{it}^{MC} such that:

$$\begin{cases} E_t\left(Z_{it}^{D}\varepsilon_{it}^{D}\right) = 0, \\ E_t\left(Z_{it}^{S}\varepsilon_{it}^{S}\right) = 0, \\ E_t\left(Z_{it}^{MC}\varepsilon_{it}^{MC}\right) = 0. \end{cases} \tag{3.13}$$

We apply the GMM to the system (Berry *et al.*, 1995). At the optimal parameter values the errors from these equations should satisfy the following moment conditions as closely as possible:

$$\begin{cases} E_t\left(Z_{it}^{D}\varepsilon_{it}^{D}\right) = \bar{m}_1 = 0, \\ E_t\left(Z_{it}^{S}\varepsilon_{it}^{S}\right) = \bar{m}_2 = 0, \\ E_t\left(Z_{it}^{MC}\varepsilon_{it}^{MC}\right) = \bar{m}_3 = 0. \end{cases} \tag{3.14}$$

The number of moment conditions exceeds the number of parameters, and therefore, following Hansen (1982), we minimize the weighted average of the squared sample moments to find the optimal parameter values:

$$\min_{\theta} \bar{m}(\theta) W^{-1} \bar{m}(\theta), \tag{3.15}$$

where $W = Asy.Var.[\bar{m}]$. The asymptotic covariance matrix of this GMM estimator is:

$$V_{GMM} = [G' W^{-1} G]^{-1} = \frac{1}{n}[G' \Psi^{-1} G]^{-1}, \tag{3.16}$$

where G is a matrix of derivatives with jth rows equal to:

$$G^{jl} = \frac{\partial \bar{m}^{l}(\theta)}{\partial \theta'} \tag{3.17}$$

and $\Psi = Var[\sqrt{n}(\bar{m} - \mu)]$.

Following Lee (1996), we first estimate the parameters by the instrumental variables technique, use these to calculate the asymptotic variance-covariance matrix, and then minimize the objective function.

3.4. Structural Estimation Results

The first system consists of inverse demand and inverse supply functions, and the FOC for profit maximization as in system (3.12). The estimates are reported in Table 3.6. The first two columns list the estimates and asymptotic standard errors for the system in which all equations are unrestricted, estimated separately for two periods. The next two columns report the estimates for a system with certain restrictions imposed on the equations.

All coefficients in the demand equations are of the expected sign. We find no statistically significant structural change in this equation between the two periods. It follows from this equation that prices are higher in the winter, in more populated and wealthier states. Implied price elasticity of demand at the mean values is found to be –3.57.

Table 3.6. Primary System, Inverse Demand/Supply Equations

Variable	**All Unrestricted**		**Partially Restricted**	
	Coefficient	**t-statistic**	**Coefficient**	**t-statistic**
Demand	1st period		Both periods	
Constant	686	2.74	1,823	8.56
Temperature, °C	−4.59	−1.38	−12.86	−4.03
Population, Mil	8.91	0.73	36.42	3.21
Income, $1000	123.8	17.44	92.8	15.21
Quantity Demanded, Bcf	−11.53	−3.87	−18.68	−6.72
	2nd period			
Constant	772	1.32		
Temperature, °C	1.61	0.18		
Population, Mil	−30.67	−0.99		
Income, $1000	98.7	8.51		
Quantity Demanded, Bcf	−4.90	−0.65		
Supply	1st period			
Constant	1,396	20.53	1,396	20.53
Reserves, Tcf	−9.78	−1.67	−9.76	−1.67
Price of Oil, $/barrel	17.4	4.97	17.4	4.97
Quantity Supplied, Bcf	0.476	0.66	0.473	0.65
	2nd period			
Constant	−763	−4.06	−764	−4.06
Reserves, Tcf	−40.58	−5.15	−40.54	−5.14

(*Continued*)

Table 3.6. (*Continued*)

Variable	All Unrestricted		Partially Restricted	
	Coefficient	t-statistic	Coefficient	t-statistic
Supply	2nd period			
Price of Oil, $/barrel	167.7	12.61	167.7	12.61
Quantity Supplied, Bcf	7.324	8.73	7.320	8.72
Transport Costs	1st period		Both periods	
Constant	998	12.02	1,040	17.05
Distance, miles 1000*s*	864.2	7.27	880.0	10.03
Distance^2	−86.65	−4.42	−84.08	−5.22
	2nd period			
Constant	1,128	12.67		
Distance, miles 1000*s*	968.9	7.37		
Distance^2	−14.36	−0.47		
Markup	1st period			
Q_{supply}, Bcf	−1.207	−0.84	−1.959	−1.39
$HHI*Q_{supply}$, Bcf	−6.312	−3.23	−7.137	−3.87
$CHHI*Q_{supply}$, Bcf	0.849	0.38	1.237	0.60
	2nd period			
Q_{supply}, Bcf	−3.999	−3.66	−2.508	−2.54
$HHI*Q_{supply}$, Bcf	−14.94	−5.18	−8.239	−3.53
$CHHI*Q_{supply}$, Bcf	−1.172	−0.76	−2.881	−2.01

The supply equation does change significantly between the two periods. The price of natural gas in the second period became more responsive to market conditions: all coefficients of the equation increase in absolute value between the periods. All signs are as expected and most coefficients are statistically significant. The price is higher for the pipeline with lower proved reserves and when the price of crude oil is higher. The coefficient on quantity supplied is insignificant for the first period. The elasticity of supply at the mean values in the second period is 5.64.

There are at least two possible explanations for the structural change in the supply equation. One is that the decline in the proportion of the long-term contracts between producers and pipelines for the firm gas supply in the second period provided producers with extra flexibility to react to changes in economic environment. Another is that the increased number of potential buyers at the wellhead market generated more favorable conditions for producers, inducing extra investment in exploration and development. Transportation cost does not change between the two periods. The marginal cost of transportation increases with the distance in a concave manner, by about 74 cents per thousand miles per thousand cubic feet of gas transported.

The markup of the FOC equation provides support for our hypothesis as to the nature of Cournot-Bertrand oligopoly. Using the estimates from the partially restricted system, we conclude that the change in CV is statistically significant. The point estimate of the coefficient of conjectural variation in the first period is 0.07, and −0.15 in the second. The interactivity of service offerings of the pipelines under full regulation changed from being more cooperative than Cournot to being less cooperative than Cournot, and more like Bertrand.

We also find support for the hypothesis that the transportation industry exhibits increasing returns to scale. The point estimates of parameters b_0 and b_1 of equation (3.8) are 12.90 and 1.22 for the first, and 13.46 and 4.91 for the second periods.

The second system consists of direct demand and supply functions in the following form:

$$\begin{aligned}\sum_{j\in\Omega_D} m_{ijt}Q^D_{jt} &= \alpha^D + \sum_{j\in\Omega_D} m_{ijt}X^D_{jt}\beta^D \\ &\quad - \sum_{j\in\Omega_D} m_{ijt}P^D_{jt}/\gamma^D + \sum_{j\in\Omega_D} m_{ijt}\varepsilon^D_{jt}, \\ Q^S_{it} &= \alpha^S + X^S_{it}\beta^S + P^S_{it}/\gamma^S + \varepsilon^S_{it}, \qquad (3.18)\end{aligned}$$

and the FOC for the profit maximization in the usual form. The estimates are reported in Table 3.7.

Table 3.7. Primary System, Direct Demand/Supply Equations

Variable	**All Unrestricted**		**Partially Restricted**	
	Coefficient	**t-statistic**	**Coefficient**	**t-statistic**
Demand	1st period		Both periods	
Constant	58.40	7.59	61.64	12.68
Temperature, °C	−0.906	−19.28	−1.039	−41.56
Population, Mil	3.302	12.80	3.684	25.94
Income, $1000	2.497	3.66	1.227	5.22
Price Demanded, $/MMcf	−0.02008	−4.47	−0.00980	−5.19
	2nd period			
Constant	68.36	8.96		
Temperature, °C	−1.161	−33.17		
Population, Mil	4.137	20.18		
Income, $1000	0.237	0.63		
Price Demanded, $/MMcf	−0.00227	−0.86		
Supply	1st period			
Constant	−11.57	−1.93	−11.92	−1.99
Reserves, Tcf	4.122	16.04	4.119	16.09
Price of Oil, $/barrel	−1.56	−8.21	−1.56	−8.21
Price Supplied, $/MMcf	0.02230	5.56	0.02245	5.60
	2nd period			
Constant	55.49	2.93	55.31	2.91
Reserves, Tcf	6.089	10.68	6.090	10.67
Price of Oil, $/barrel	−12.00	−6.49	−11.98	−6.48
Price Supplied, $/MMcf	0.06898	8.96	0.06887	8.93

We estimate both restricted and unrestricted versions of the equation. The coefficients in the demand equation are of the expected sign. Quantities demanded of natural gas are higher in the winter, in more populated and wealthier states. Implied price elasticity of demand at the mean values is −0.65 for the restricted estimation, and −1.34 for the unrestricted estimation for the first period. The coefficient on price for the second period is insignificantly different from zero.

Again, the supply equation changes significantly between the two periods. The quantity supplied of natural gas in the second period becomes more responsive to market conditions: all the coefficients of the equation increase in absolute value between the periods. All signs are as expected and all coefficients are statistically significant. The quantity supplied is higher for the pipeline with access to more proved reserves, and when the price of crude oil is lower. The elasticity of supply at the mean values increases from 0.92 in the first period to 2.85 in the second.

Estimated transportation costs and markup parameters do not change significantly compared to those reported in Table 3.6, and therefore, are not reported in Table 3.7.

Third, we estimate the system that accounts for fixed effects at the pipeline and state level in the demand and supply equations:

$$\sum_{j\in\Omega_D} m_{ijt} Q^D_{jt} = \alpha^D_j + \sum_{j\in\Omega_D} m_{ijt} X^D_{jt}\beta^D - \sum_{j\in\Omega_D} m_{ijt} P^D_{jt}/\gamma^D + \sum_{j\in\Omega_D} m_{ijt}\varepsilon^D_{jt},$$

$$Q^S_{it} = \alpha^S_i + X^S_{it}\beta^S + P^S_{it}/\gamma^S + \varepsilon^S_{it}, \qquad (3.19)$$

where α^D_j and α^S_i are now allowed to take state and pipeline specific values. Introduction of the fixed effects accounts for most of the variance in Population, Income, and Reserves variables. Therefore, to avoid multicollinearity, we remove these from the analysis. Instead, we put in temperature squared as an extra explanatory variable on the demand side.

The results are reported in Table 3.8. We do not report state or pipeline fixed effect coefficients here. We expect to find a negative relation between temperature and quantity demanded[16]: this is the case, with the coefficient highly statistically significant. The implied elasticities of demand are −0.24 and −0.08 for the two periods. No structural change occurred in the demand equation in the two periods.

The supply equation did change. Again, for the first period, the coefficient on the price supplied is both of the wrong sign and statistically insignificant. For the second

[16]We presume that gas use is not dominated by air conditioning.

Table 3.8. State/Pipeline Fixed Effects

Variable	**All Unrestricted**		**Partially Restricted**	
	Coefficient	**t-statistic**	**Coefficient**	**t-statistic**
Demand	1st period		Both periods	
Temperature, °C	−3.213	−23.63	−3.385	−38.47
Temperature^2	0.02033	17.99	0.02173	28.59
Price Demanded, $/MMcf	−0.003630	−1.42	0.000994	0.85
	2nd period			
Temperature, °C	−3.402	−24.83		
Temperature^2	0.02228	17.97		
Price Demanded, $/MMcf	−0.001134	−0.60		
Supply	1st period			
Price of Oil, $/barrel	−1.03	−6.87	−1.03	−6.87
Price Supplied, $/MMcf	−0.00324	−1.19	−0.00310	−1.14
	2nd period			
Price of Oil, $/barrel	−14.02	−8.45	−14.01	−8.44
Price Supplied, $/MMcf	0.08833	11.56	0.08829	11.56

(*Continued*)

Table 3.8. (*Continued*)

Variable	**All Unrestricted**		**Partially Restricted**	
	Coefficient	**t-statistic**	**Coefficient**	**t-statistic**
Transport Costs	1st period		Both periods	
Constant	242.2	13.31	242.5	18.51
Distance, miles, 1000*s*	216.9	7.72	207.2	9.91
Distance^2	−28.11	−5.73	−23.56	−6.07
	2nd period			
Constant	247.8	12.97		
Distance, miles, 1000*s*	195.0	6.21		
Distance^2	−12.22	−1.83		
Markup	1st period			
Q_{supply}, Bcf	2.976	2.19	3.216	2.47
$HHI{*}Q_{\text{supply}}$, Bcf	−6.506	−4.02	−6.734	−4.40
$CHHI{*}Q_{\text{supply}}$, Bcf	−0.721	−0.38	−0.831	−0.47
	2nd period			
Q_{supply}, Bcf	−1.389	−1.41	−1.577	−1.84
$HHI{*}Q_{\text{supply}}$, Bcf	−5.819	−2.92	−3.918	−2.22
$CHHI{*}Q_{\text{supply}}$, Bcf	−1.101	−0.97	−1.689	−1.57

period, the coefficient has the correct sign: supply becomes more responsive to the economic conditions, with an implied elasticity at the mean values of 3.65. The equation asserts that higher prices of crude oil induce lower quantities supplied of natural gas. Point estimates of the conjectural variation coefficient decline over the two periods, but the change is statistically insignificant, and the demand slope is insignificantly different from zero.

Next, Table 3.9 reports the estimation in which the prices and quantities are measured by the logarithm of their values. Thus, coefficients on the quantities show the price elasticities.

According to these estimates, the quantity demanded is higher in the winter, in more populated and wealthier states. The price elasticity of demand in the first period is –0.57, insignificantly positive for the second period, and insignificantly negative for the two combined.

The supply equation changes in the second period. The elasticity of the quantity supplied with respect to the price of crude oil is decreasing from –0.06 in the first period to –0.23 in the second. *Ceteris paribus*, pipelines that have more reserve backing for throughput provide more gas to the market. The elasticity of supply changes from a statistically insignificant 0.18 in the first period to 2.79 in the second.

Table 3.9. Logarithm of Prices and Quantities

Variable	All Unrestricted		Partially Restricted	
	Coefficient	**t-statistic**	**Coefficient**	**t-statistic**
Demand	1st period		Both periods	
Constant	7.202	4.60	4.190	4.71
Temperature, °C	−0.02991	−36.93	−0.03020	−59.22
Population, Mil	0.07716	16.42	0.07760	25.78
Per Capita Income, $1000	0.06941	5.50	0.03791	6.78
$\ln(P_{Demand})$, ln($/MMcf)	−0.5683	−2.53	−0.0986	−0.82
	2nd period			
Constant	3.356	2.90		
Temperature, °C	−0.02887	−41.24		
Population, Mil	0.07258	16.53		
Per Capita Income, $1000	0.03482	3.51		
$\ln(P_{Demand})$, ln($/MMcf)	0.0136	0.08		
Supply	1st period			
Constant	2.14	1.24	2.14	1.24
Reserves, Tcf	0.1211	13.76	0.1212	13.77
ln(Price of Oil), ln($/barrel)	−0.0636	−8.48	−0.0636	−8.48
$\ln(P_{Supply})$, ln($/MMcf)	0.180	0.74	0.180	0.74
	2nd period			
Constant	−14.92	−7.14	−14.93	−7.14
Reserves, Tcf	0.1398	7.36	0.1398	7.36

(*Continued*)

Table 3.9. (*Continued*)

Variable	**All Unrestricted**		**Partially Restricted**	
	Coefficient	**t-statistic**	**Coefficient**	**t-statistic**
Supply	2nd period			
ln(Price of Oil), ln($/barrel)	−0.2264	−4.91	−0.2264	−4.91
ln(P_{Supply}), ln($/MMcf)	2.793	8.03	2.793	8.03
Transport Costs	1st period		Both periods	
Constant	997	12.01	1,041	17.07
Distance, miles 1000*s*	864.2	7.27	880.2	10.04
Distance^2	−86.67	−4.42	−84.06	−5.22
	2nd period			
Constant	1,128	12.67		
Distance, miles 1000*s*	969.2	7.38		
Distance^2	−14.29	−0.47		
Markup	1st period			
Q_{Supply}, Bcf	−1.209	−0.84	−1.957	−1.38
$HHI*Q_{Supply}$, Bcf	−6.313	−3.23	−7.136	−3.87
$CHHI*Q_{Supply}$, Bcf	0.852	0.38	1.236	0.60
	2nd period			
Q_{Supply}, Bcf	−3.998	−3.66	−2.502	−2.54
$HHI*Q_{Supply}$, Bcf	−14.944	−5.19	−8.237	−3.53
$CHHI*Q_{Supply}$, Bcf	−1.172	−0.76	−2.886	−2.01

Transportation costs increase with distance in a concave manner, by about 76 cents per thousand miles per thousand cubic feet of gas transported. The conjectural variation parameter decreases in the second period.

3.4.1. *Estimated elasticities*

Table 3.10 summarizes the estimates of demand and supply price elasticities for the systems described previously.

The estimates of the elasticity of supply for the second period are consistent across various specifications of the model, all being statistically significant, and falling in a range between 2.80 and 3.65. The estimates for the first

Table 3.10. Elasticities at the Mean Values

System	**Demand**			**Supply**	
	1st period	**2nd period**	**Both periods**	**1st period**	**2nd period**
Primary	***−1.33	−0.15	***−0.65	***0.93	***2.85
Fixed Effects	−0.24	−0.08	0.07	−0.13	***3.65
Logarithms	***−0.56	0.02	−0.10	0.18	***2.80

***Statistically significant estimates.

period are much smaller. One out of three estimates for the first period statistically differs from zero.

The estimates of the demand elasticities are spread out across the range of values. Estimates range from 0.07 to –1.33, with three estimates statistically significant with point estimates at –0.56, –0.65, and –1.33. The estimates for the second period are smaller in absolute values, and statistically insignificant from zero. The estimates of the demand and supply elasticities at mean values are comparable across various specifications of the model equations.

The structural system resolves the inconsistency of the reduced form estimates. Primarily, these findings on the effect of deregulation on demand and supply quantities are that they are due to a structural change in the supply equation for the deregulation period that has not been explicitly accounted for in the reduced form approach. The model also has allowed us to separate the changes in the degree of competition from the other changes in the supply side of the market.

3.5. Evaluations of the Past and Future Policies

Here we would like to conduct a few policy evaluation exercises. In the first, we want to use the estimates from the structural model to evaluate the effects of Orders 636 *et al.*

on consumer and producer surpluses, and on transportation industry profits.

The explicit use of the pipeline architecture in the model provides a unique tool for the evaluation of new pipeline construction projects. These are added to the existing network, with consequent changes in prices and quantities calculated. As an example, in the second policy analysis, we forecast the general impact that the construction of a proposed Alaskan pipeline would have on existing prices and quantities. The construction of such a pipeline has been suggested as a possible way to meet increasing natural gas demands in the continental United States and Canada caused in part by higher crude oil prices.

3.5.1. *The welfare effects of the Orders 636 et al.*

It follows from the reduced form regressions of the basis differential that the average price difference between the price of gas at the wellhead and the price of gas at the city gate, or the transportation charge, declined in the second period by $171 per million cubic feet of gas transported. To find the total decline in annual transportation charges, we multiply this number by the total amount of gas transported by all interstate pipelines (11.2 Tcf in 1997), and arrive at an estimate of \$1.93 billion. Indeed, the total savings to consumers might be even greater, because we do not account for the decline in intrastate transportation rates, which follow the same pattern. If instead we multiply the decline in charges by the total US consumption of gas (about 20 Tcf per year), we would get an estimated \$3.42 billon in annual savings.

Alternatively, we can estimate this effect of restructuring the contract markets using the structural model estimates. We employ the values from Table 3.6 for this purpose. The total per period profit that a pipeline collects is given by the difference between the total revenue in the form:

$$TR_{it} = \sum_{j \in \Omega_D} m_{ijt} P^D_{jt} Q^S_{it} \tag{3.20}$$

and the total costs:

$$\begin{aligned} TC_{it} = \sum_{j \in \Omega_D} m_{ijt}(c_{ij} + \varepsilon_{ijt} \\ - b_0 m_{ijt} Q^S_{it} - b_1 Q^S_{it} + P^D_{jt}) Q^S_{it}. \end{aligned} \tag{3.21}$$

It follows from the FOC (3.10) that:

$$\begin{aligned} TR_{it} - TC_{it} = ((\gamma^D - b_0) HHI^+_{it} \lambda \gamma^D \\ CHHI^+_{it} \gamma^S - b_1)(Q^S_{it})^2. \end{aligned} \tag{3.22}$$

Therefore, we estimate the annual savings to consumers from the decline in transportation charges as

$$\sum_{t=1}^{12}\sum_{i\in\Omega_S}(Q_{i,1997}^{S})^2 \begin{pmatrix} [(\gamma^D - b_0)_{\mathrm{II}} - (\gamma^D - b_0)_{\mathrm{I}}]HHI_{i,1997} \\ +(\gamma^S - b_1)_{\mathrm{II}} - (\gamma^S - b_1)_{\mathrm{I}} \\ +[(\lambda\gamma^D)_{\mathrm{II}} - (\lambda\gamma^D)_{\mathrm{I}}]CHHI_{i,1997} \end{pmatrix} \quad (3.23)$$

where t stands for the month, *I* and *II* for the first and second regulatory period.

The total decline is found to be \$1.05 billion. This corresponds to an average decline of about \$93 per MMcf transported (about 7.2 percent), with \$19 per MMcf decline due to the change in the transportation costs parameter b_0, \$59 per MMcf increase mainly due to the change in the elasticity of supply, and \$133 per MMcf decline due to the change in the competitive environment. These estimates are slightly below the ones derived from the reduced form regressions.

3.5.2. *The market impact of a proposed Alaska pipeline into the Mid-North American region*

For this analysis, we take estimates of demand and supply from the primary system of direct demand/supply equations as reported in Table 3.7. Demand is taken for the entire time interval, and supply for the period after deregulation. Due to the fact that transportation costs and markup equation coefficients differ quite significantly for the primary and fixed effects systems, we take an average of the two sets of estimates to form a third forecast equation.

According to a statement made by Deputy Secretary of Energy T. J. Glauthier, gas reserves in the Alaska North Slope currently available for exploration are estimated at levels of 32–38 trillion cubic feet while the reserves near Prudhoe Bay are estimated at 22 trillion cubic feet.[17] Canada's National Energy Board also projected that about 9 trillion cubic feet have been discovered in the Mackenzie River Delta and Beaufort Sea, with a further 55 to 75 trillion still waiting to be found. To be conservative, we use currently discovered reserves only of 30 trillion cubic feet as the amount to be transported to the mid-Atlantic region of the United States over a number of years.

The distance from the reserves to the consumption market is calibrated at 2,900 miles from Prudhoe Bay, AK to

[17]T. J. Glauthier, Deputy Secretary, US Department of Energy, "Testimony to the Senate Committee on Energy and Natural Resources" (September 14, 2000).

Chicago, IL. Due to the remoteness of Alaska from the interior of the continent, any other specific origin or delivery point does not change the distance significantly.

The price of oil that is used is the average for the twelve months September 1999–August 2000 with domestic first purchase crude oil ($24.19 per barrel) measured in real 1990 dollars ($21.52 per barrel). These prices are much lower than those experienced in 2004–2007, but we take them to be approximate to long run contract dollar levels.

Another variable that must be calibrated before we are able to proceed with the forecasting is the state markets to which the new pipeline will be connected, and in what state-by-state proportions it will pump gas. Let us consider two scenarios here. First, let S_{new} be proportional to the year 1997 total annual deliveries to each state, i.e.,

$$S_{\text{new}} = \frac{1}{\sum_{j \in \Omega_D} Q^D_{j,1997}} \begin{bmatrix} Q^D_{1,1997} \\ Q^D_{2,1997} \\ \cdots \\ Q^D_{J,1997} \end{bmatrix}, \tag{3.24}$$

where S_{new} denotes the shares of the new pipeline in the consuming states, and $Q^D_{j,97}$ denotes the total gas deliveries to the state j in 1997. Second, assume that the new pipeline delivers gas equally to all states, i.e.,

$$S_{\text{new}} = \begin{bmatrix} 1/N_S \\ 1/N_S \\ \dots \\ 1/N_S \end{bmatrix}, \tag{3.25}$$

where N_S is a total number of such states.

The system to assess market impact then consists of the demand equations at the state level, the supply equations at the wellhead of the existing and new pipeline, the FOC of profit maximization for the exiting and new pipeline, and the adjusted transportation identity that accounts for a new source of supply. The system is exactly identified and, therefore, has a unique solution.

According to the model, for the first scenario of proportional shares the price of gas at the Alaska wellhead will be \$1.65 per thousand cubic feet; the quantity supplied by the new pipeline to all states will be 94.1 billion cubic feet per month, or about 1.13 trillion cubic feet per year. The number corresponds to about 5 percent of the current consumption levels. The other pipelines will experience an average price decline at the wellhead of 5.8 cents per Mcf, or an approximately 3.4 percent decrease from current levels. The average price decline in the consuming states will be about 3.9 cents per Mcf, which corresponds to a 1.3 percent decrease. This corresponds to approximately \$0.78

billion in annual consumers' savings. Due to the increased pressure from the north, the existing pipelines will cut their total supply by about 83.4 Bcf per month. Consumers in all states will experience increased consumption at lower prices, with a total increase of 10.7 Bcf per month.

Figure 3.1 shows the estimated changes for the individual states. Consumers in Massachusetts, Minnesota, New Jersey, New York, and Wisconsin will benefit the most, with increased consumption of more than 1 Bcf per month. Arkansas, Iowa, Nebraska, Ohio and Virginia will face the lower supplies due to general price declines at the wellhead, and increased competition that the pipelines

Figure 3.1. Change in Quantity Demanded, by State, Proportional Shares, Bcf/month

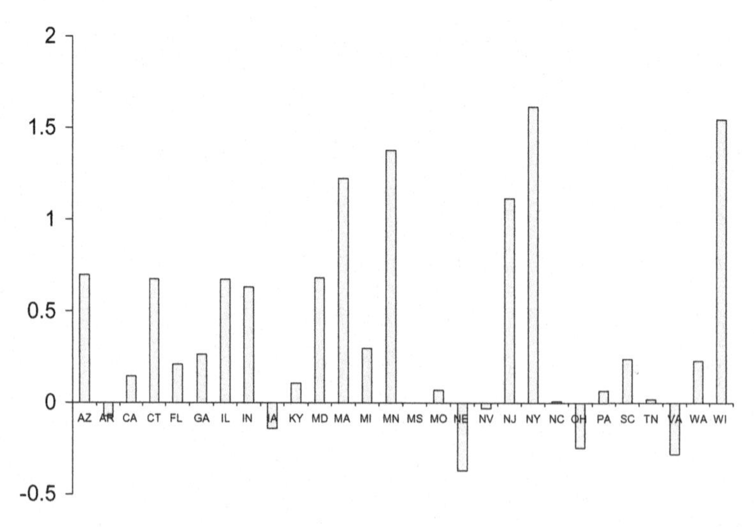

delivering gas there will have to fight in other states. This last observation is a bit unrealistic due to the fact that the model does not allow pipelines to readjust their transportation shares and change the proportion of deliveries to the states. It is indeed logical that new construction will cause the existing pipelines to reevaluate the profitability of corresponding markets, and to readjust their deliveries in some fashion.

The prices at the wellhead of the other pipelines will be suppressed due to the new Alaskan supply shifting out the aggregate supply curve. Individual effects are depicted in Figure 3.2. Although it is not true for all large pipelines in

Figure 3.2. Change in Wellhead Prices, by Pipeline, Proportional Shares, cents/Mcf

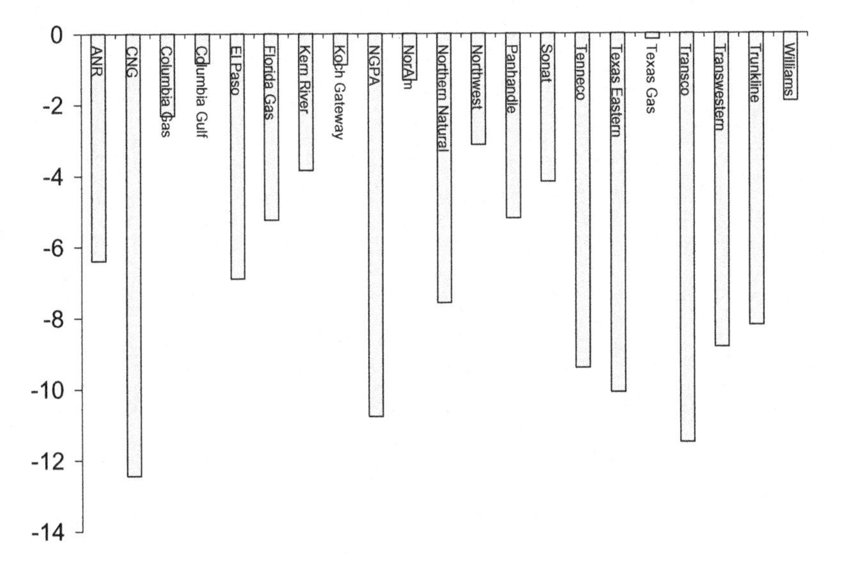

the sample, most of them will have to adjust purchasing, or wellhead, prices downward to meet the competitive pressure. CNG, NGPA, Texas Eastern, and Transco will have to reduce prices by more than 10 cents per Mcf; Northern Natural, Tenneco, Transwestern, and Trunkline will pay 8 to 9 cents less to their respective gas providers. Smaller pipelines, such as Columbia Gas, Columbia Gulf, Koch Gateway, NorAm, Texas Gas, and Williams, that face more concentrated markets, and are therefore, able to respond more aggressively to the changes, would not experience such a dramatic decline.

According to the model estimated for the pipeline with equal deliveries to all states, the price of gas at the Alaska wellhead will be $1.48 per Mcf; the quantity supplied by the pipeline to all states will be 81.9 Bcf per month, or just short of 1 Mcf per year. That number corresponds to 4–5 percent of the current consumption levels. The other pipelines will experience an average price decline at the wellhead of 4.1 cents per Mcf, or an approximately 2.4 percent decrease from current levels. The average price decline in the consuming states will be about 8.1 cents per Tcf, which corresponds to a 2.7 percent decrease. This corresponds to approximately $1.62 billion in annual consumers' savings. Due to the increased pressure from the north, the existing pipelines will cut their supply by about 59.8 Bcf per month. Consumers will experience increased consumption at lower prices, with a total increase of 22.1 Bcf per month.

As for individual states, consumers in smaller states will indeed benefit more, because according to the assumption of equal shares of the new pipeline the supply to such states would increase relatively more than for average and larger states. The largest hit in the wellhead price falls on the Northern Natural Gas Pipeline, of about 10 cents per Mcf; as for the rest, the decline is pretty evenly distributed among them.

3.6. Conclusions

3.6.1. *Summary and applications*

This chapter introduces and estimates an integrated demand and supply model of the natural gas industry that explicitly accounts for existing pipeline network architecture. The structural model takes into account geographic separation of supply and demand markets. The existing interactions between markets that are linked by a pipeline are used to generate otherwise unavailable variables for model identification.

The structure of the network has not been explicitly used in previous published research. The results show that accounting for network architecture and natural gas flows provides new estimation results. It allows for full utilization of the cross-sectional variation of the data. The model includes behavioral variables that identify the extent of interpipeline Cournot and Bertrand pricing in the market, and permits us to assess the effect of the recent regulatory changes on that behavioral pattern. The results are robust across various specifications of demand and supply. Controlling for possible structural change in the demand and supply equations, we find that the competition in the partial deregulation period increased, as measured by the change in the average conjectural variation in the market from 0.07 to –0.15.

The model also reveals significant change in the supply equations in the post-regulation period not reported before in the literature. Reduced form regressions show a decline in transportation charges by 12 percent in the post-regulation period. Total annual savings to consumers are estimated at $1.93 billion per year. A corresponding estimate from the structural model is $1.05 billion per year.

The model is further applied to analyze the short-run impact of the construction of a new pipeline on natural gas prices and quantities. A hypothetical pipeline from the Alaska North Slope is chosen as an example. The resulting impact on prices and quantities that such a pipeline would bring about is calculated, and it is concluded that such a pipeline would provide supply that reduces prices and thereby results in between \$0.78 and \$1.62 billion in annual consumers' savings at the city-gate.

3.7. Appendix

3.7.1. *Data transformation example*

Let us illustrate a procedure that we employed to define various variables by an example. Let a simple network consists of three states, labeled S_A, S_B, and S_C, and two pipelines that deliver gas to these states, labeled P_K and P_L. Let annual realized deliveries be given by the following transportation matrix:

$$T = \begin{array}{c|ccc} & S_A & S_B & S_C \\ \hline P_K & 10 & 20 & 10 \\ P_L & 30 & 30 & 40 \end{array}. \tag{3.26}$$

According to this matrix, pipeline P_K transports 10 units to states S_A and S_C, and 20 units to state S_B. The total volume that P_K moves to all states is 40 units, which is a sum of the corresponding row of the transportation matrix; the total volume delivered to state S_A is 40 units, which is a sum of the corresponding column of T; and similarly for other pipelines and states. Let us denote a vector of quantities delivered to states by Q^D (for "quantity demanded"), and a vector of quantities shipped by pipelines by Q^S (for "quantity supplied"). In this example,

$$Q^D = \begin{bmatrix} Q_A^D \\ Q_B^D \\ Q_C^D \end{bmatrix} = \begin{bmatrix} 40 \\ 50 \\ 50 \end{bmatrix} \tag{3.27}$$

and

$$Q^S = \begin{bmatrix} Q_K^S \\ Q_L^S \end{bmatrix} = \begin{bmatrix} 40 \\ 100 \end{bmatrix}. \tag{3.28}$$

From the transportation matrix, we then determine the pipelines' shares in each state. We call this new matrix a *state normalized* transportation matrix, T_S. To find the first column of T_S, we take the first column of T and divide each element by the total deliveries to state S_A, or the sum of the first column of T. In our example, we get:

$$T_S = \begin{array}{c|ccc} & S_A & S_B & S_C \\ \hline P_K & 0.25 & 0.40 & 0.20. \\ P_L & 0.75 & 0.60 & 0.80 \end{array} \tag{3.29}$$

Similarly, we define a *pipeline normalized* transportation matrix, T_P, as the original transportation matrix, element-by-element divided by total shipments through each pipeline, or the sum of the rows of T:

$$T_P = \begin{array}{c|ccc} & S_A & S_B & S_C \\ \hline P_K & 0.25 & 0.50 & 0.25. \\ P_L & 0.30 & 0.30 & 0.40 \end{array} \qquad (3.30)$$

Notice the important properties of these matrices that follow from the construction: the columns of T_S and the rows of T_P are always summed to one. Also notice that

$$Q^D = T_P' \cdot Q^S \qquad (3.31)$$

and

$$Q^S = T_S \cdot Q^D. \qquad (3.32)$$

Substituting one into the other, we get:

$$Q^D = T_P' \cdot T_S \cdot Q^D. \qquad (3.33)$$

However, $T_P' \cdot T_S$ is not an identity matrix. In fact, there exist infinitely many matrices that satisfy equation (3.33) in place of $T_P' \cdot T_S$. Also note that $X \neq T_P' \cdot T_S \cdot X$ for an arbitrary vector X.

Suppose we are given annual transportation matrix T and monthly deliveries to each state Q_t^D. We assume that pipelines' shares in the states are constant throughout the year, and defined by T_S. It then follows that we can find monthly shipments by pipeline Q_t^S using equation (3.32):

$$Q_t^S = T_S \cdot Q_t^D. \tag{3.34}$$

That way, we calculated monthly quantities that each pipeline transported, and used these values in our estimation procedures.

Next, suppose that we are given some demand shifter, X^D, defined on a state level:

$$X^D = \begin{bmatrix} X_A^D \\ X_B^D \\ X_C^D \end{bmatrix}. \tag{3.35}$$

Then, to find a value of the variable at the pipeline level, we project it into the pipeline space by the following matrix operation:

$$X_S^D = T_P \cdot X^D = \begin{bmatrix} 0.25X_A^D + 0.50X_B^D + 0.25X_C^D \\ 0.30X_A^D + 0.30X_B^D + 0.40X_C^D \end{bmatrix}. \tag{3.36}$$

Similarly, the variable defined at the pipeline level, X^S, can be projected into state space:

$$X_D^S = T_S' \cdot X^S = \begin{bmatrix} 0.25X_K^S + 0.75X_L^S \\ 0.40X_K^S + 0.60X_L^S \\ 0.20X_K^S + 0.80X_L^S \end{bmatrix}. \tag{3.37}$$

Notice that, this operation may be repeated twice, and applying $T_P \cdot T_S'$ or $T_S' \cdot T_P$ to a vector would, in general, result in a different vector.

What is the economic interpretation of the i-th component of the vector X_S^D? It is the average value of X^D in the market that is defined as relevant for pipeline i. What is the interpretation of the j-th component of the vector $T_S' \cdot X_S^D$? It is the weighted average value of X_S^D in the state j market, or the average value of the variable X_S^D for the relevant competitors of state j (also including state j).

CHAPTER 4

Quantitative Study Number Three

THE BASIS DIFFERENTIALS ON PARTIALLY DEREGULATED PIPELINE TRANSPORTATION

Paul W. MacAvoy

The basis differential between two hubs in the pipeline network is the difference between the prices of spot gas sold in markets at the two points. When these markets are integrated the basis differential is made up of the differences in the current spot prices of gas at the origin and destination hubs which should equal the pipeline charge for transportation of that gas from the origin to the upstream point of delivery.

If this gas is sold at that destination hub, its price there equals the sum of these two. That is, the prices of the commodity between the two hubs should not differ by more than the cost (price) of transportation between the points; if it did, then arbitrage in the market would result in higher entry and/or lower exit gas prices until the difference shrinks to the transportation cost. Alternatively, a deregulated gas transportation market, in which line charges were flexible with respect to space demand changes, would clear at transport price levels equal to the basis differential.

Thus, basis differentials during the period following partial deregulation provide an indication of how pipeline performance was affected by policy change. Prices on contracts for firm commitments of pipeline space were not deregulated, but were capped at average historical costs of a unit of space. As a result, the demand-induced sharp increases in basis differentials could not feedback to pipeline pricing.

4.1. Methodology

Three historical data sets were obtained from Natural Gas Intelligence, the daily, weekly, and monthly (bidweek) natural gas index prices at about eighty-five trade points in the US and Canada. The monthly historical data cover the years from 1986 to 2004, while the weekly and daily data set start from 1996 and extend to 2004. Since indexes are established at the different hubs on different dates, the data start at different dates for each of the hubs. It is assumed that dates with zero represent either no trade or lack of data.

The monthly (bidweek) gas price indexes are used to estimate basis differentials as they allow for capturing the trends for a longer time series as compared to the daily prices that are available only from 1996. Chapter 2 makes the case to delineate the pipeline activity into market regions — Northeast, Central or Midwest, and Western. We consider only differentials between the major field source hubs in Texas, Canada, and the Rocky Mountains and the three destination regions. First, we constructed a table of hub pairs of origins and destinations in four sets. Set A consisted of city pairs with the destinations in the Northeast; Set B of pairs terminating in the major hubs in the Midwest; Set C of city pairs with destination hubs in the Western region with California as focus; Set D made up city pairs of "unrealistic" basis differentials, in which there is no physical flow between the two points, but this set is essential as a control on the preceding sets.

The basis differential is obtained by subtracting the price at the origin hub from that at the destination hub. For sets A, B, and C the differential cell is empty whenever any of the following conditions applies: index at the origin is greater than at destination, index is missing or is zero at either the origin or destination. However, due to the nature of the investigation, for set D, the differential cell is empty only when the index is nil in either the origin hub or the destination. As such, negative values appear in those calculations and are seen in the figures.

For ease of graphical comparison, the horizontal and vertical scale for the figures of sets A, B, and C have been made the same. As with A, B, and C the horizontal scale for D is similar spanning the period from December 1989 to December 2003, however, the Y-axis alters to accommodate a different scale of the basis differential.

4.2. Observations

The figures show that, between 1996 and 2003, there were six distinct spikes in thc basis differentials originating from the Southwest hubs and terminating at the hubs in the Northeast. Destination point Transco Zone Six recorded the highest of any of those spikes. Though of different degrees, they tended to occur at the same time across several Northeast hubs suggesting the existence of a shortage of pipeline capacity in the regional market within the Northeast.

Within the Midwest, there is also an indication of a common market from the common spikes. However, the spikes are much smaller in scale than those in the Northeast.

Table 4.1. Principal Pipelines By Market Region

Company Name	Principal Market*	System Capacity (MMcf/d)**
Algonquin Gas Transmission Company	**Northeast**	**2,154**
Carnigie Natural Gas Co	Northeast	40
Columbia Gas Transmission Corporation	**Northeast**	**7,276**
Cove Point LNG LP	**Northeast**	**1,000**
Dominion Transmission, Inc	**Northeast**	**6,275**
East Tennessee Natural Gas Company	Northeast	738
Eastern Shore Natural Gas Company	Northeast	111
Equitrans, LP	Northeast	822
Granite State Gas Transmission, Inc.	Northeast	150
IPOC as Agent/Iroquois Gas Trans. Sys., LP	**Northeast**	**850**
Maritimes & Northeast Pipeline,c LLC	Northeast	440
National Fuel Gas Supply Corporation	Northeast	2,168
Portland Natural Gas Transmission System	Northeast	178
St Lawrence Gas Co	Northeast	62
Tennessee Gas Pipeline Company	**Northeast**	**7,271**
Texas Eastern Transmission, LP	**Northeast**	**6,438**
Transcontinental Gas Pipe Line Corporation	**Northeast**	**7,362**
Vermont Gas System Inc	Northeast	49
Alliance Pipeline LP	**Midwest**	**1,767**
ANR Pipeline Company	**Midwest**	**6,667**
Centra Pipelines Minnesota Inc	Midwest	63
Crossroads Pipeline Company	Midwest	250
Great Lakes Gas Transmission Limited Ptrshp	Midwest	2,895
Guardian Pipeline, LLC	Midwest	750
Horizon Pipeline Company, LLC	Midwest	380
Midwestern Gas Transmission Company	**Midwest**	**748**
Mississippi River Transmission Corp	Midwest	1,670
Natural Gas Pipeline Company of America	**Midwest**	**5,001**
Northern Border Pipeline Company	**Midwest**	**3,094**
Northern Natural Gas Company	**Midwest**	**3,904**
Panhandle Eastern Pipe Line Company	**Midwest**	**2,765**
Trunkline Gas Co	Midwest	1,884
Viking Gas Transmission Co	Midwest	543
El Paso Natural Gas Company	**Western**	**4,882**
Kern River Gas Transmission Company	**Western**	**898**
Northwest Pipeline Corporation	**Western**	**2,900**
Paiute Pipeline Company	Western	160
PG&E Gas Transmission, Northwest Corporation	**Western**	**2,700**
Transwestern Pipeline Company	**Western**	**2,836**
Tuscarora Gas Transmission Company	Western	120
Mojave Pipeline Co	Western	400

Source: EIA, Expansion and Change on the US Natural Gas

However, the spikes could have been a result of a domino effect on the demand side from the large Northeast bottlenecks, but the relationship lacks the strength to suggest that it, and Midwest, are in the same market (see Chapter 2).

The focused spikes in California make it clear that the West was not within the same market boundary as the rest of the country. The magnitudes of the observed spikes suggest that capacity bottlenecks that contributed to the spikes of 2001 there were the greatest then elsewhere experienced.

4.3. Basis Differential Spikes by Hub and Spoke Networks

As described in Chapter 2, three or four separate pipeline networks have overlapped each other to traverse a quadrant of the country, e.g., the Northeast from Virginia border to the upper tier of New England and New York State. The Transcontinental, Tennessee Gas, Texas Eastern, and Columbia systems take most of the gas to the Northeast from the large field reserves in Louisiana, onshore and offshore, and East Texas through "pipeline alley" along the Coast (again from both older fields onshore and the newer larger deposits in State and Federal producing blocks offshore). These volumes are then traded by producers to retail utilities, industrial buyers or their agents mostly at the five entry hubs shown in Table 4.2 "Set A." They are shipped on "firm"

Table 4.2. City Pairs for Bidweek Basis Differential Calculation

Origin	**Destination**
SET A : Terminating at North East Market Centers	
1. Henry Hub	Dominion Hub
	Transco Zone 6, New York
	TETCO, New Jersey
	Algonquin Citygate
	Iroquois Center, New York
	NE Columbia Gas
	Tennessee, Zone 6
2. Katy Hub, Texas	TETCO, New Jersey
	Tennessee Zone 6
	Transco Zone 6, New York
3. Carthage Hub, Texas	Dominion Hub
	Transco Zone 6
	Iroquois Centre, New York
	TETCO, New Jersey
4. Texas Eastern, E. Louisiana	Transco Zone 6, New York
	Dominion Hub
	NE Columbia Gas
	TETCO, New Jersey

(*Continued*)

Table 4.2. (*Continued*)

Origin	Destination
SET B: Terminating in the Midwest Consumption Centers	
1. Henry Hub	Chicago City Gate
	ANR Joliet Hub
	Michigan Consolidated
2. Katy Hub	Chicago City Gate
	ANR Joliet Hub
3. Carthage Hub	Chicago Citygate
4. Waha Hub	Chicago Citygate
	ANR Joliet Hub
	Michigan Consolidated
5. Mid-Continent (Oneok Gas), KS	Chicago Citygate
	ANR Joliet Hub
6. Cheyenne/Colorado Interstate Gas, CO	Chicago Citygate
7. Mid-Continent NGPL	Chicago Citygate
8. El Paso Permian, West Texas	Chicago Citygate
SET C: Terminating in Western Hubs (California)	
1. Transwestern Pipeline, West Texas	Southern California Border Average
	Southern Border, PG&E
	Southern Border, SOCAL
	PG&E Citygate

(*Continued*)

Table 4.2. (*Continued*)

Origin	**Destination**
2. El Paso Permian , West Texas	PG&E Citygate
	Southern California Border Average
	Southern Border, PG&E
	Southern Border, SOCAL
3. Opal, WY	Southern Border, PG&E
	PG&E Citygate
	Southern Border, SOCAL
	Southern California Border Average
4. Kingsgate Center on US/Canada Border	PG&E Citygate
SET D: City Pairs without physical connections	
PG&E Citygate	Chicago Citygate
	Transco Zone 6, NY
Opal Hub, WY	Dominion Hub, PA
	Chicago Citygate
CIG, Colorado	TETCO, New Jersey
El Paso Permian, TX	NE Columbia Gas
Southern California Border Average	Dominion Hub
	Chicago Citygate
Henry Hub, LA	Southern California Border Average

or "interruptible" space contracts in these pipelines to destination hubs also as shown in Table 4.2 in New Jersey or New York State (except for Columbia and Dominion, which distribute volumes in Virginia or surrounding locations).

Figures 4.1 and 4.2, for basis differentials on links between two entry nodes to the Columbia gas exit node, indicate substantial variation at or below $0.50 per MMBtu for delivery costs. Most of this variation was seasonal, with differentials at $0.25 per MMBtu in the period from March to September when pipelines were operating substantially below full capacity. When winter heating demands for gas were at their peak, the differential reached $0.50 between September and March; by inference, the bidweek prices for transportation did not reach the regulatory caps on firm

Figure 4.1. Bidweek Basis Different b/w Texas Eastern E. LA and NE Columbia Gas, US$/MMBtu

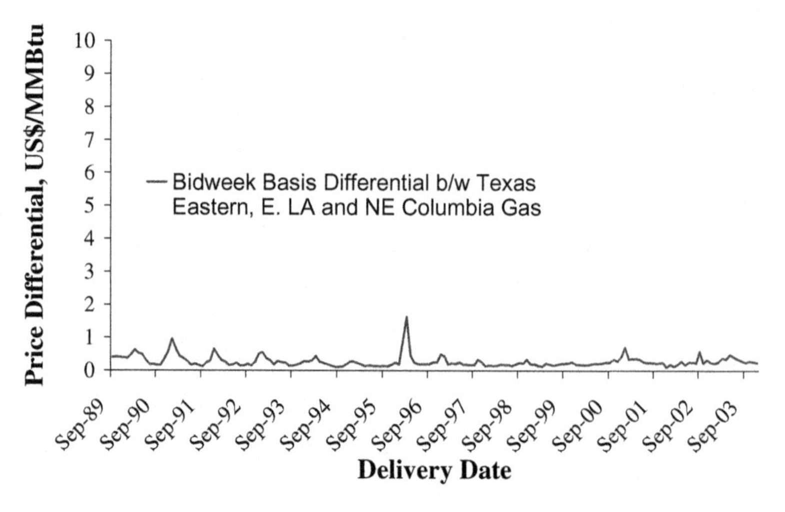

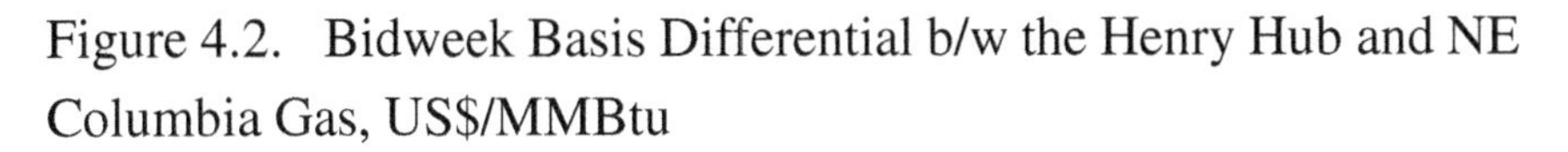

Figure 4.2. Bidweek Basis Differential b/w the Henry Hub and NE Columbia Gas, US$/MMBtu

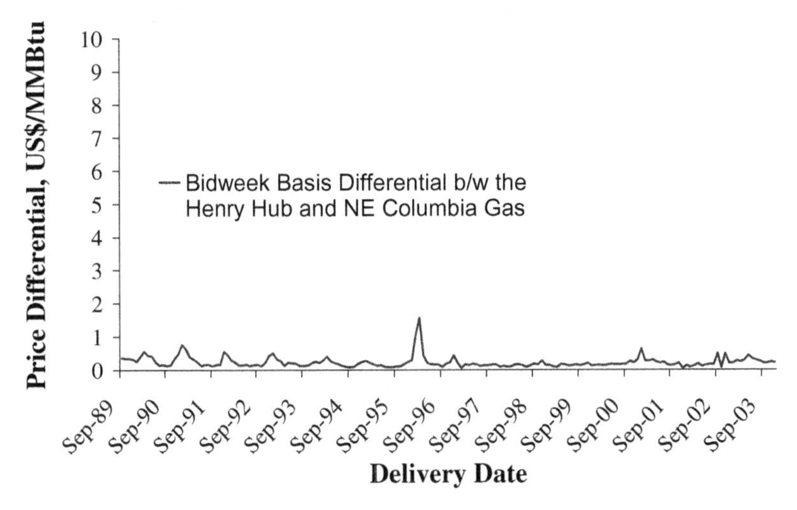

transportation in the summer by half, but were at these caps in the winter.

There were one or two quarters, in the winter of 1996, when the differentials exceeded the caps, increasing to \$1.50 per MMBtu. In effect the prices in the exit hub at Columbia set by retail consumers, industrial users, and brokers were bid for the full capacity throughput to three times the transport cap price level. This took place on throughput from both the Texas Eastern entry hub and the Henry hub, so that there had to have been a market-wide "shortage" at the exit hub for gas to be delivered by the Columbia distribution system. The "shortage" could only have been excess demand for space in the existing nodes of pipeline systems to the Columbia exit node, to be "worked" through the market by increased buyer prices for the gas that was available at that exit node.

Figures 4.3 and 4.4 document the basis differentials on the shipments between the (Texas Eastern) Henry entry hubs and the Dominion exit hub in Pennsylvania. Here, the implied prices off-season were scarcely $0.25 MMBtu, but on-season were in excess of $0.50 in at least 10 quarters. The price spike between September 1995 and March 1996 was approximately $2.00 per MMBtu, $0.50 greater than at the Columbia exit hub. This spike was repeated between September and March 2003; if price caps on throughput to Dominion were in the $1.00 MMBtu range, then markets for spot gas were cleared at twice the regulated price caps.

Figure 4.3. Bidweek Basis Differential b/w the Henry Hub and the Dominion Hub, Pennsylvania, US$/MMBtu

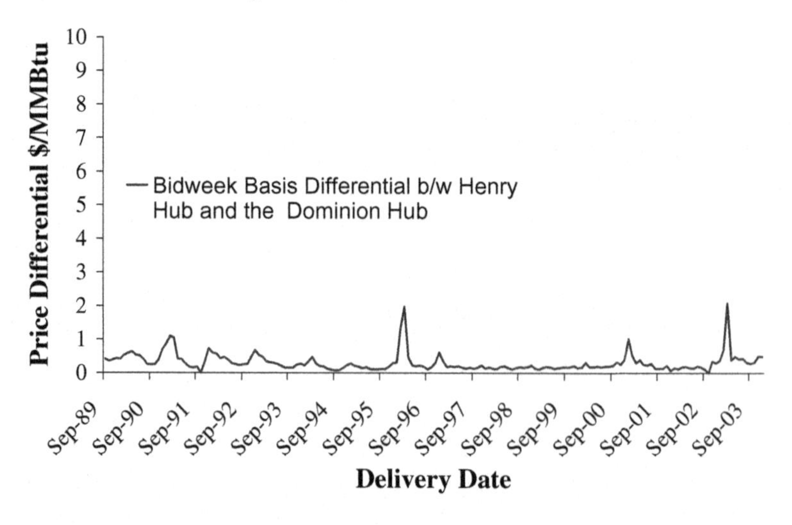

Figure 4.4. Bidweek Basis Differential b/w Texas Eastern, E. LA and Dominion Hub, US$/MMBtu

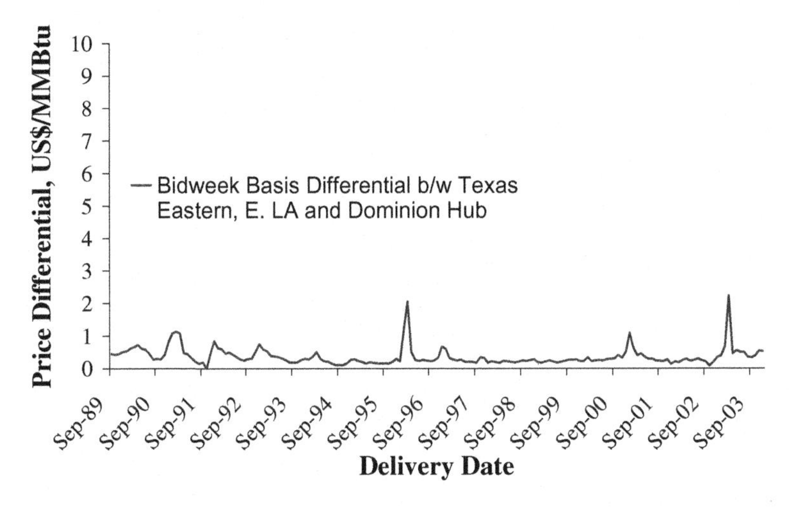

The New Jersey exit nodes of the major pipelines serve as the dominant spot markets for gas to the Northeast industrial and population centers of the country. The basis differentials for gas shipped on the second largest of the three major pipelines, Texas Eastern Transmission Company, are shown in Figures 4.5 and 4.6. Data were available for only the period after September 1995; they indicate caps on prices for pipeline space in the range of \$0.75 – \$1.00 per MMBtu, achieved by bidweek charges in five of the on-season quarters between 1996 and 2003. The rest of these on-season periods were impacted by differential spikes ranging from \$2.50 to \$4.25 MMBtu. Excess demands for spot gas at the TETCO exit hub, due to lack of pipeline capacity, were eliminated by bids at exit hubs

Figure 4.5. Bidweek Basis Basis Differential b/w Henry Hub and TETCO, NJ, US$/MMBtu

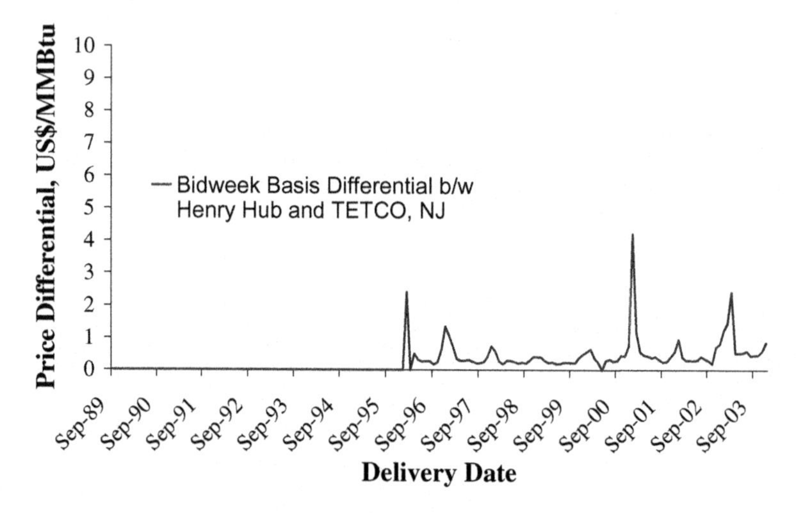

Figure 4.6. Bidweek Basis Differential between Katy Hub, TX, and TETCO (NJ), US$/MMBtu

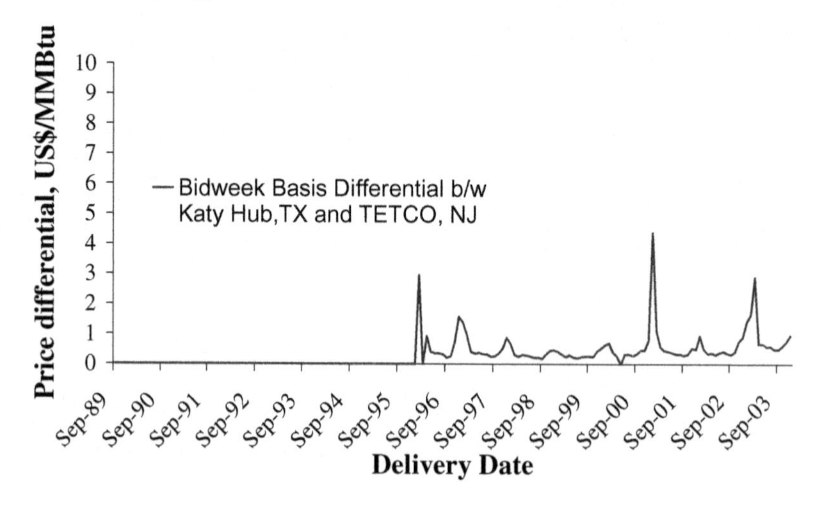

two to four times the transportation caps. As many as eight on-season markets were destabilized by lack of capacity to realize all demands for space at the TETCO hub.

The instability of pricing under caps in the largest pipeline service market is shown by the basis differentials between four entry hubs in Texas and Louisiana, and the two exit hubs Transco and Transco Six in New York. The data series are truncated to include only the five years covering the off-seasons of 1998 to 2003. All have four basis differential spikes in the five on-seasons, three in the range of $2.50 to $3.00 per MMBtu. The spike of the winter from October 2000 to March 2001 was marked by differentials averaging $9.00 MMBtu, implying that exit hub

Figure 4.7. Bidweek Basis Differential b/w Henry Hub, LA and Transco Zone 6, NY, US$/MMBtu

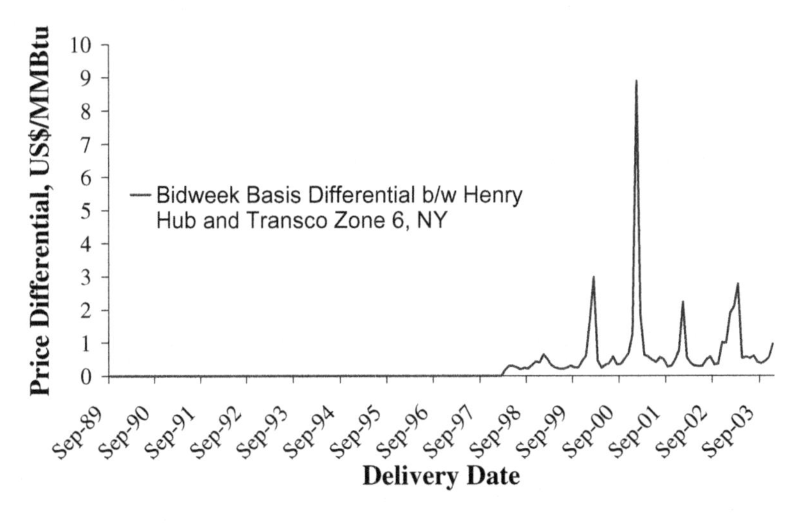

Figure 4.8. Bidweek Basis Differential b/w Katy Hub, TX and Transco Zone 6, NY, US$/MMBtu

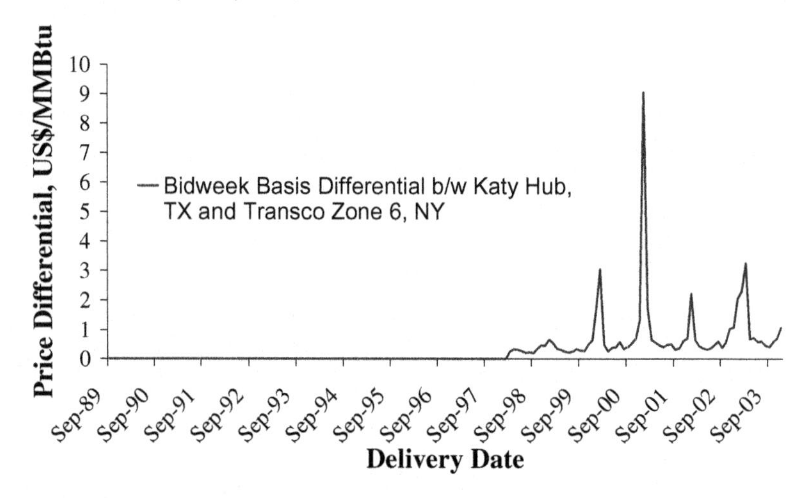

Figure 4.9. Bidweek Basis Differential b/w Carthage Hub, TX and Transco Zone 6, NY, US$/MMBtu

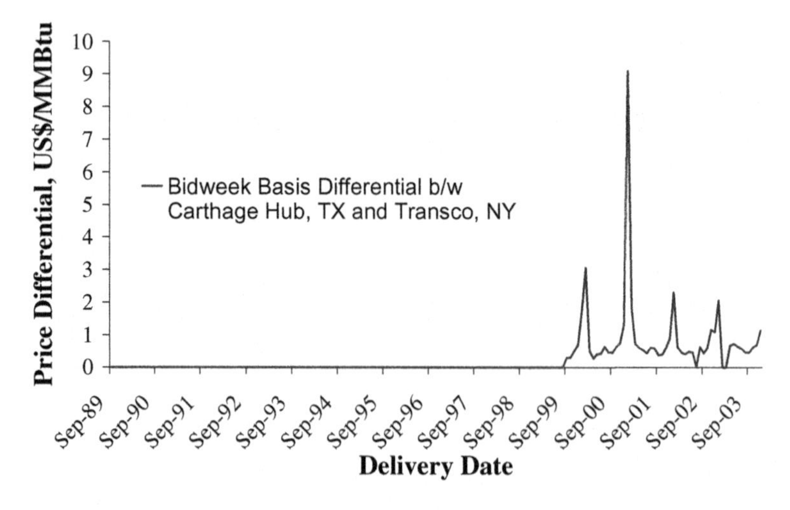

spot prices for gas exceeded entry spot prices plus transport caps by at least $7.50 per MMBtu. The largest pipeline system, that of Transcontinental gas was short significant

Figure 4.10. Bidweek Basis Differential b/w Texas Eastern, E. LA and Transco Zone 6, NY, US$/MMBtu

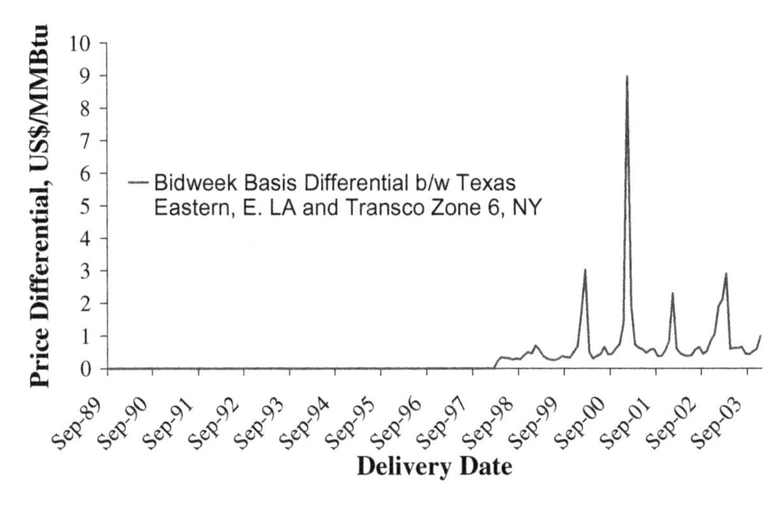

supplies of gas needed to clear final consumers demands in New England, and only by trebling the spot price at Transco Zone Six could this excess demand be choked off.

4.4. Pipeline Basis Differential Performance in the Midwest and Western Transportation Markets

With the growth of retail demands falling short of that in the Northeast, the Midwest markets did not experience the spikes in wholesale spot prices at exit hubs to the same extent. A representative basis differential series is shown in Figure 4.11, for gas transported from the Henry Hub to the pipeline exit hub for Michigan Consolidated. From 1992 to 2003 the off-season basis differentials seldom exceeded

Figure 4.11. Bidweek Basis Differential b/w Henry Hub and the Michigan Consolidated, US$/MMBtu

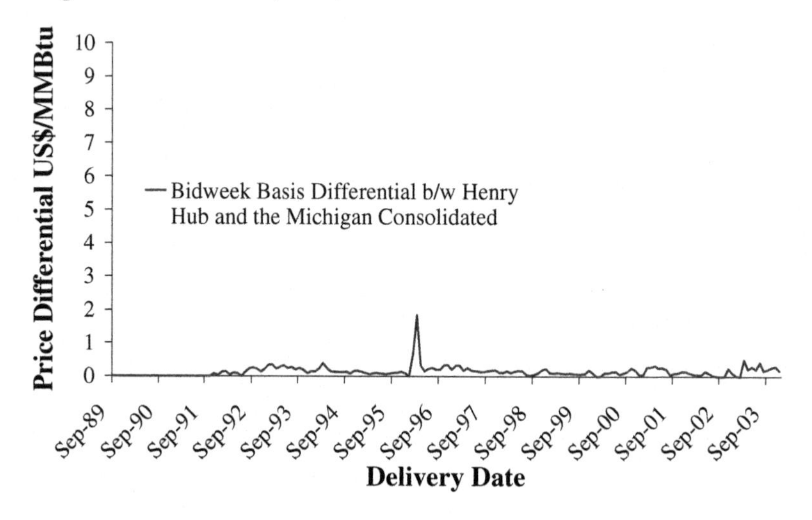

$0.25, and the on season were less than $0.50 MMBtu, indicating transport prices short of caps greater than $0.50 MMBtu. The sharp increase in retail demands experienced in the on-season of 1996 in the Northeast was also experienced in the Midwest, resulting in a spike at the full line capacity in gas at the exit hub just short of $2.00 MMBtu. But there were no basis differential spikes in the early 2000s, just an increased tendency on-season for the differentials to reach $0.50 MMBtu.

The center of gas exit in the Midwest has been the Chicago hub, with links to not only the Texas entry hubs but also the Rocky Mountain and most recently the Canadian entry hub at Kingsgate. The Chicago Citygate experienced the same basis differential behavior as Michigan Consolidated, except there was a limited spike in differentials in the winter of 2000–2001 on transport from Henry, the Mid-Continent Center, and Katy (Texas) hubs (as shown in Figures 4.12–4.15). While there are differences in the length of the data series, the basis differentials all were at the level of $1.00 MMBtu, at or above the caps that applied for different pipelines and distances. Only on the pipeline transfers at intermediate hub, for gas from Waha in West Texas for shipment to Michigan Consolidated were basis differentials off-track. As shown in Figure 4.16, transactions on spot gas did not generate differentials until 1998. They did not result in a spike in differentials until 2001, but in 2003 (see Figures 4.17 and 4.18), a $2.00 MMBtu

Figure 4.12. Bidweek Basis Differential b/w the Henry Hub and ANR Joliet Hub, US$/MMBtu

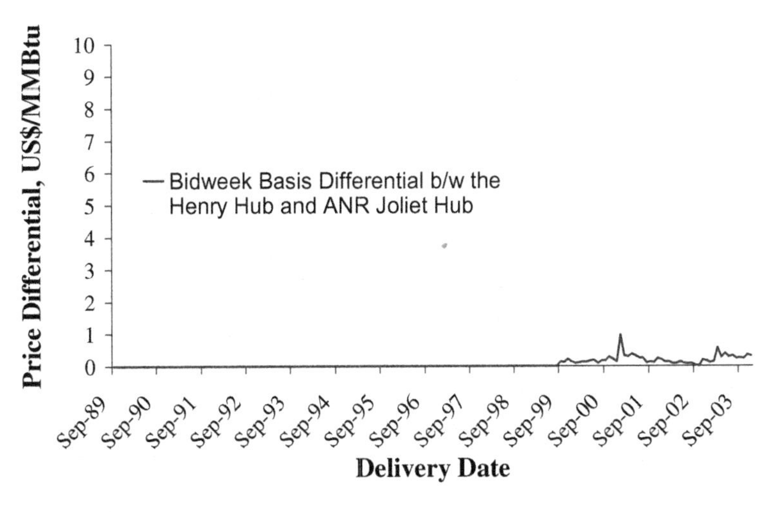

Figure 4.13. Bidweek Basis Differential b/w Mid-Continent Center, KS and Chicago Citygate, US$/MMBtu

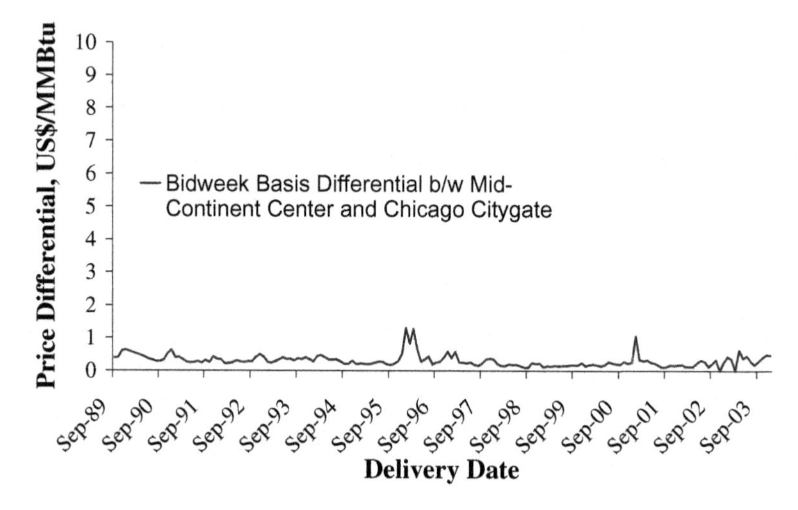

Figure 4.14. Bidweek Basis Differential b/w Katy Hub, TX and Chicago Citygate, US$/MMBtu

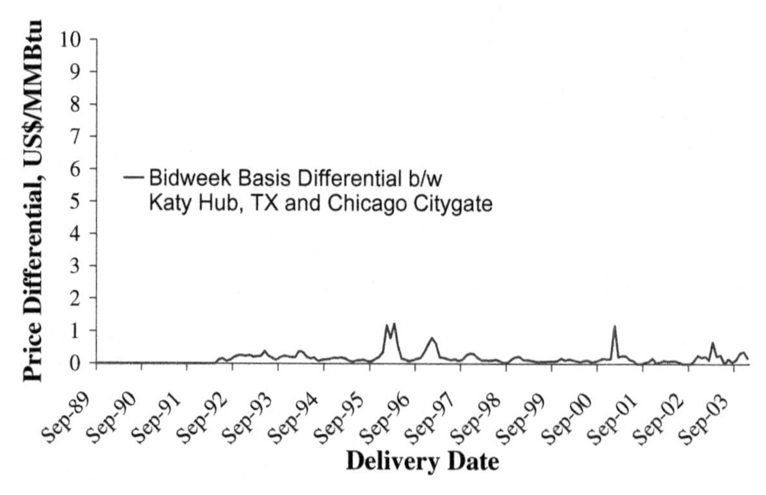

Figure 4.15. Bidweek Basis Differential b/w Mid-Continent NGPL and Chicago Citygate, US$/MMBtu

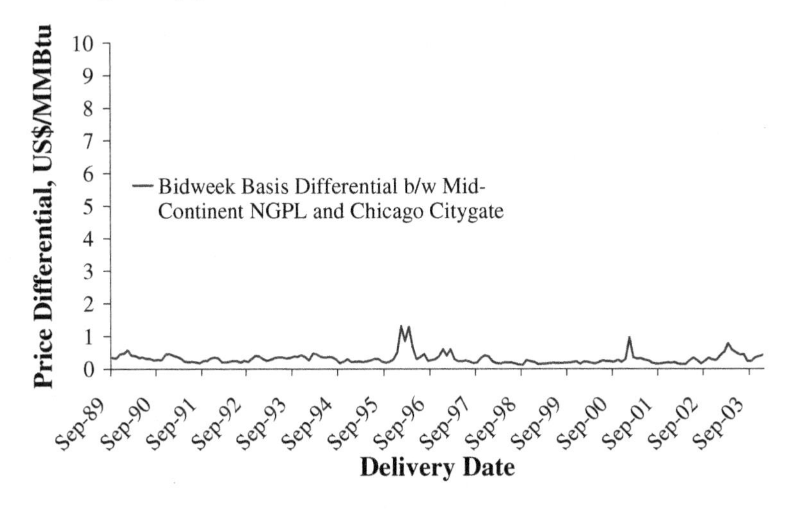

Figure 4.16. Bidweek Basis Differential b/w Waha Hub, TX and Michigan Consolidated, US$/MMBtu

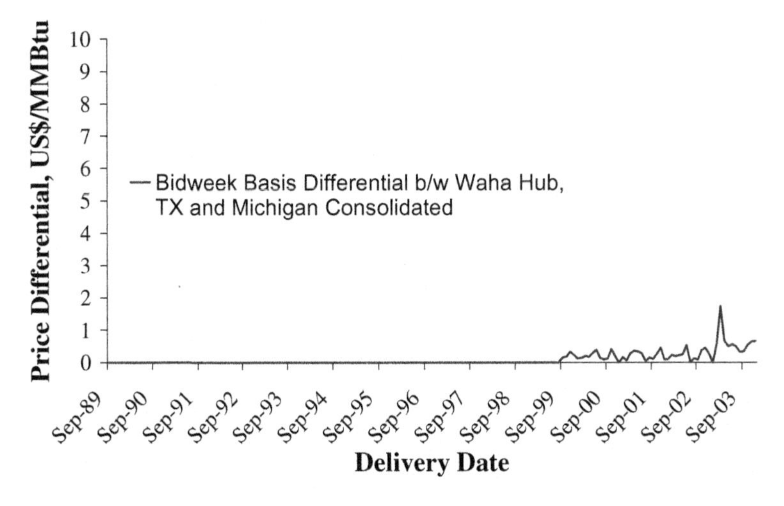

Figure 4.17. Bidweek Basis Differential b/w Waha Hub, TX and ANR Joliet Hub, US$/MMBtu

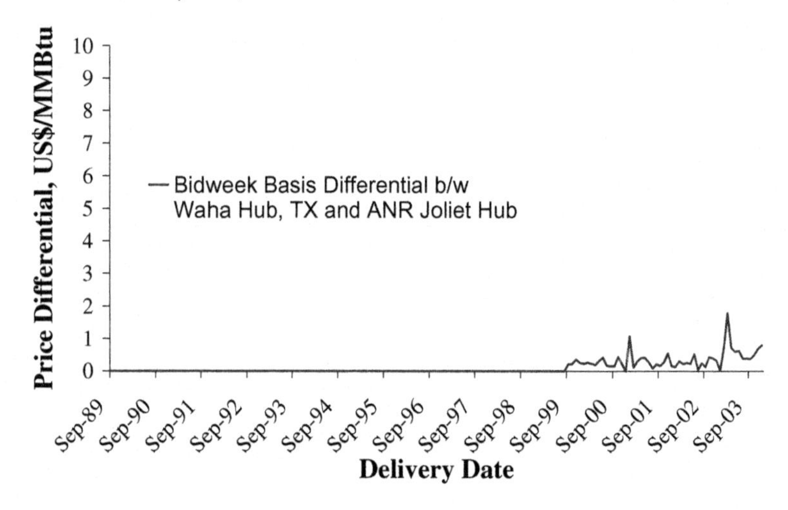

Figure 4.18. Bidweek Basis Differential b/w Waha Hub, TX and Chicago Citygate, US$/MMBtu

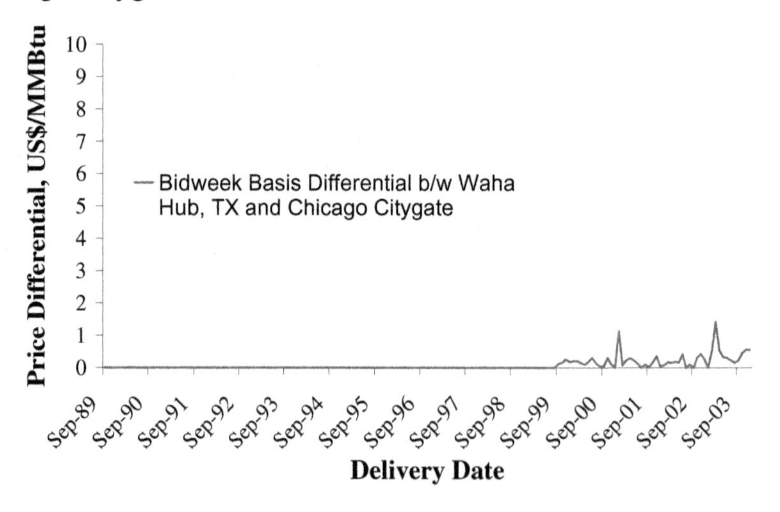

spike was realized concurrently on pipeline shipments to Chicago and Joliet from Waha; the bottleneck in capacity from West Texas had led to a doubling of spot price bids at these exit points for gas from these source hubs.

The gas and transport source most recent to join the Chicago hub market is based on the Eastern slope of the Rocky Mountains, from reserves not developed until the late 1980s. When connected by substantial links in the 1990s, prices at the wellhead and at the Chicago hub differed by $0.50 off-season and as much as $0.75 to $1.00 on-season. But during the spikes in exit hub prices at Chicago, the differentials for Rocky Mountains CIG were greater, $2.00 in 1996, $2.50 in 2001, and up to $4.50 per MMBtu over the extended period from off-season to the winter of 2002–2003 (see Figure 4.19). The explanation must center on a lack of capacity to expand deliveries from the newly-developing reserves with the lowest wellhead prices.

The western gas transmission market, since deregulation, has experienced the most significant demand shocks in spot gas transactions at receptor hubs. As a consequence of having the most severe limits on line capacity, the basis differentials there increased tenfold. Consider first the bidweek basis differential between entry hub spot gas in the Permian Basin fields of West Texas and exit hubs on the Southern California border. As shown in Figure 4.20,

Figure 4.19. Bidweek Basis Differential b/w Rocky Mts, CIG and Chicago Citygate, US$/MMBtu

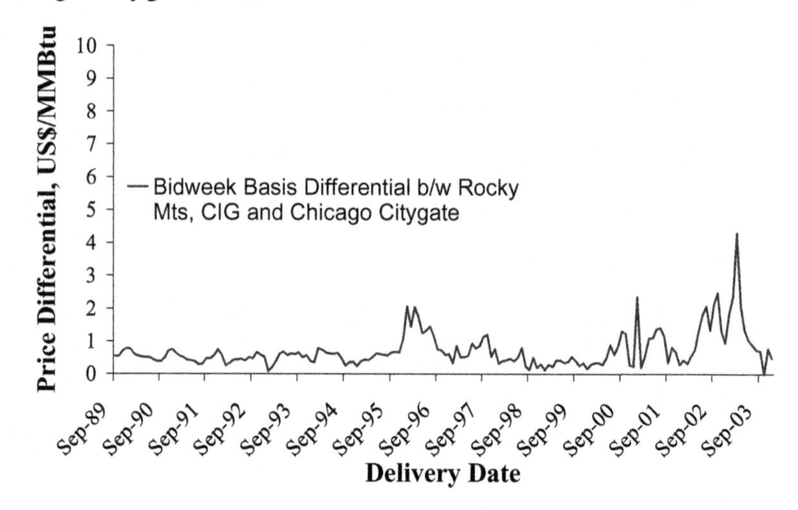

Figure 4.20. Bidweek Basis Differential b/w El Paso Permian Waha Hub, TX and Southern CA Border Ave, US$/MMBtu

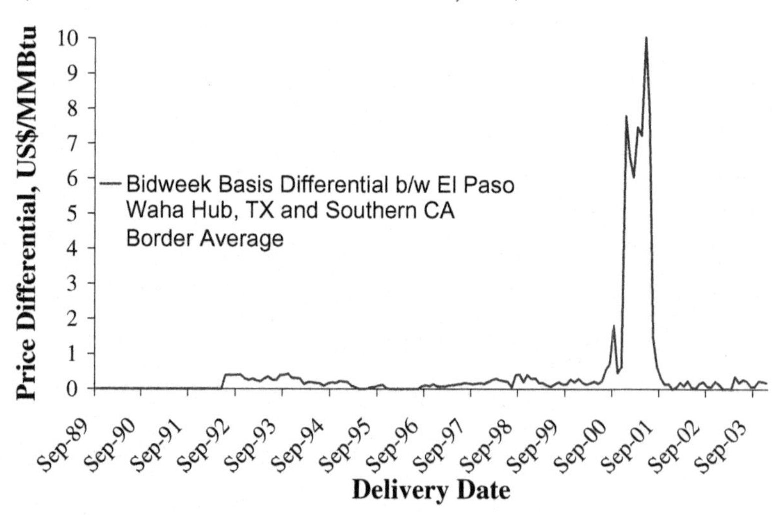

this differential varied between $0.50 per MMBtu in the peak demand periods in California and less than $0.10 per MMBtu during off-peak demand (when the oldest pipelines from Texas were close to empty). In the winter of 2000, the differential increased to $1.50, then to $7.50 per MMBtu. Early in 2001 it fell to $6.00, but increased the next month again to $7.50 and then in March 2001 to $10.00 per MMBtu (see Figure 4.20). The price for market clearing spot gas on the Southern California Border included a premium over the wellhead price of the equivalent of $60 per barrel of crude oil.

This was not unique to buyers from those Permian Basin sources that shipped gas over the El Paso gas transmission

Figure 4.21. Bidweek Basis Differential b/w El Paso Permian, West TX and Southern California Border Average, PG&E, US$/Btu

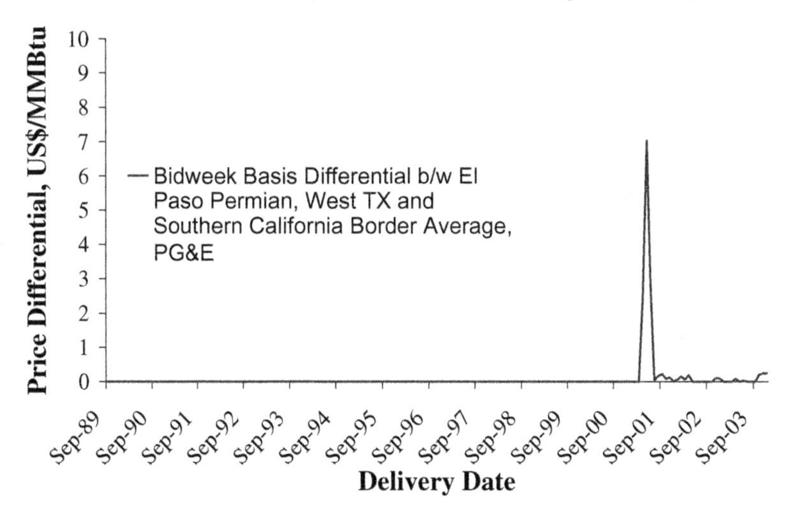

systems. Pacific Gas and Electric Company, seeking spot gas for retail distribution throughout Northern California, paid almost the full amount of this premium in the same period. After years of dispute with El Paso, PG&E restarted taking substantial shipments from El Paso at its receptor hubs, to experience in 2001 this basis differential spike (of $7.00 per MMBtu; see Figure 4.23). On gas not from the Permian Basin Fields but from the spot markets at the West Texas Waha hub, PG&E experienced successive spikes of $7.50, $6.00, and $7.40 per MMBtu between September 2000 and March 2001 (see Figure 4.22). On gas shipped over the Transwestern pipeline, both to the Southern California border and to PG&E's exit hubs, the basis differential

Figure 4.22. Bidweek Basis Differential b/w El Paso Permian, Waha Hub, TX and PG&E Citygate, CA, US$/MMBtu

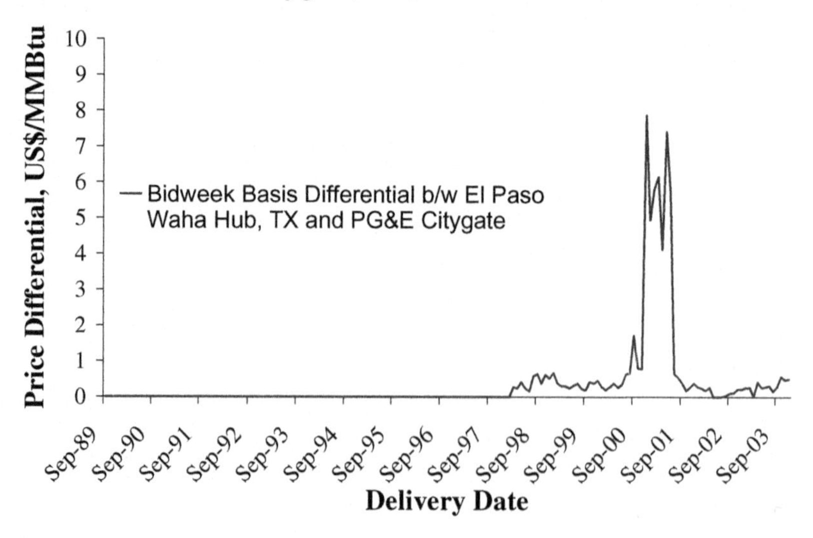

spikes were exactly the same as for service on the El Paso system — initially in the $2.00 range, increasing to $7.50 and then to $10.00 per MMBtu between September 2000 and March 2004 (see Figures 4.23 and 4.24)

The survey of basis differentials, for other entry-exit hub pairs in the Western market, indicates that the entire market for gas into California was in a bottleneck of excess demands that could not be matched by additional throughput in the pipelines from Texas, or the Rocky Mountains, or from Western Canada. All three reserve sources had the capacity to provide two-thirds of California's gas demands in a normal heating peak period. But none could carry that amount through the existing pipeline space.

Figure 4.23. Bidweek Basis Diff b/w Transwestern (TWP), West TX and Southern California Border Average, US$/MMBtu

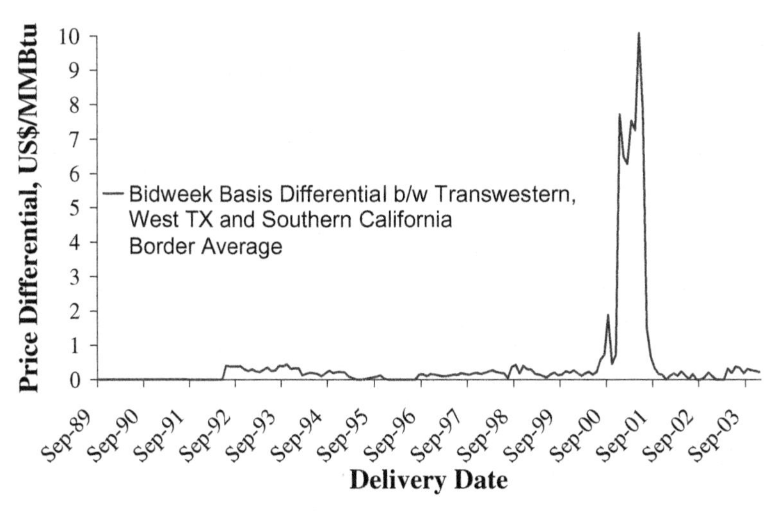

Figure 4.24. Bidweek Basis Diff b/w Transwestern (TWP), West TX and PG&E Citygate, US$/MMBtu

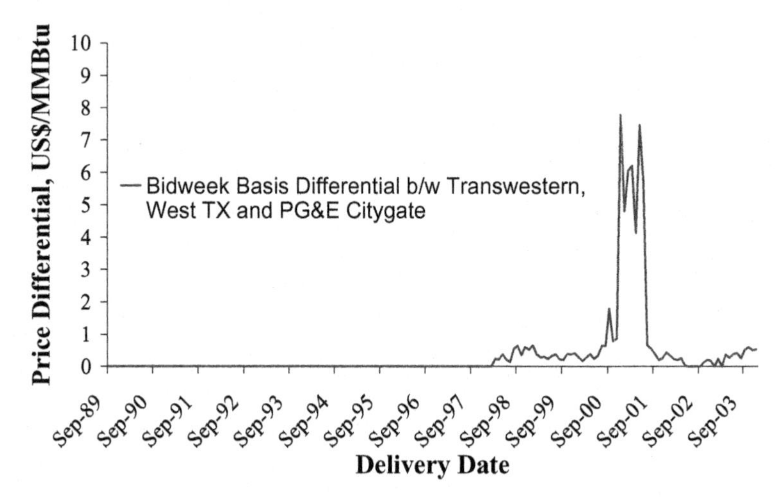

Figure 4.25 documents the basis differentials on gas source and transport from the newer gas fields on the Western slopes of the Rocky Mountain range. With price data available only after March 2001, the estimated differential begins at $8.00 per MMBtu and declines to the $0.50 range immediately, only to reestablish in the $1.50 range until March 2003. Quite similar differentials were realized for Opal Wyoming to PG&E citygate hubs except the congestion occurred on service in September 1999 forward, with differential spikes of $2.50 in September 2000, of $8.00 in October 2000 and again in May 2001 (see Figure 4.26). On service between Kingsgate Center and the PG&E hub for retail distribution, the number of spikes was higher, but

Figure 4.25. Bidweek Basis Differential b/w Opal, WY and Southern CA Border, PG&E, US$/MMBtu

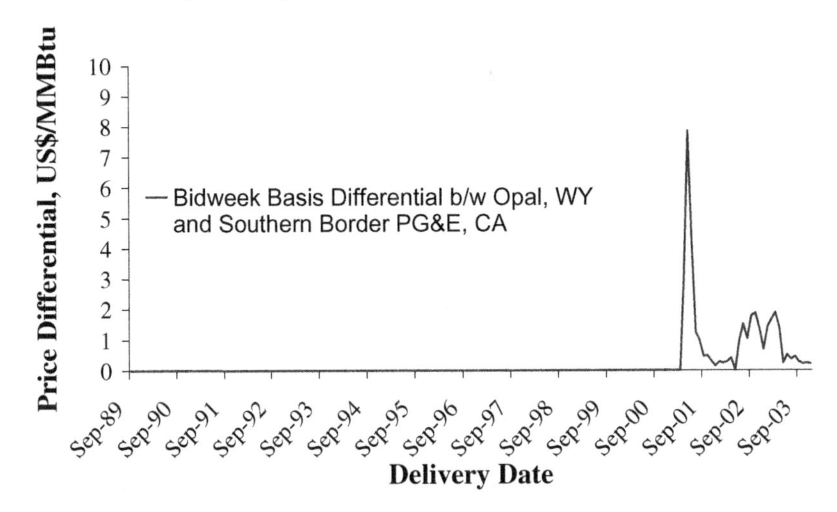

Figure 4.26. Bidweek Basis Differential b/w Opal Hub, WY and PG&E Citygate, US$/MMBtu

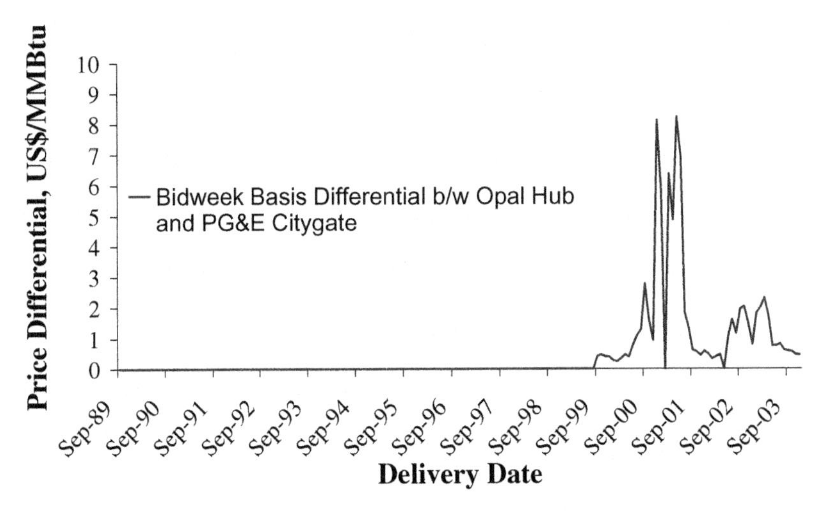

Figure 4.27. Bidweek Basis Differential b/w Kingsgate Center, ID and PG&E Citygate, CA, US$/MMBtu

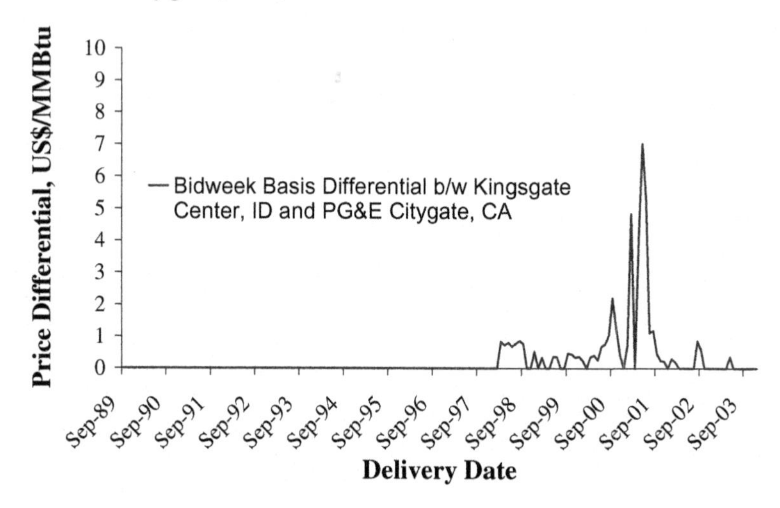

the \$8.00 differential spike was the same as for gas to be transmitted from all other sources to California (see Figure 4.27).

4.5. Basis Differentials and Pipeline Throughput

Then what are the implications for performance of the pipeline networks in the post Order 636 period? The basis differential spikes did originate in exceptional and unexpected increases in demand for pipeline space, of the order $Q_t = (Q_{t-1} + \Delta Q_t)$ where Q_t is realized utilization of space in year t and is equal to last periods utilization Q_{t-1} plus the further amount to take the line to full capacity ΔQ_t. But with a surge in demand, this increases to the level $Q_t + dQ_t$ where d is some fraction of realized space demand that is unrealized, because there is no further unsold line capacity. Performance can be evaluated by the size of the losses to shippers, not from having to pay the price-cap price P for line capacity that is used, but from having to pay ΔP more to clear out the excess demand dQ_t (the amount $Q_t \Delta P$ on all shipments subject to price adjustments due to the increase in the current auction market at peak), as well as having to go without the shipments that would constitute use of capacity beyond current full capacity (the dollar value of foregone user surplus $dQ_t \times \Delta P/2$).

Losses of these amounts must have been very large, as indicated by even the crudest of approximations. Transco's full capacity throughput from the main lines from the Henry Hub to New York – New Jersey Zone Six for the peak winter months in 2000–2009 was approximately 250 Bcf/day and the excess demand for space generating the basis spike of January 2001 was in the range of 20 percent, so that dQ_t equaled 50×30 Bcf for that month while ΔP, the spike, was at least \$8.00 per Mcf or \$8 million per Bcf. The deadweight loss was \$6 billion in the month of January, from just one set of lines of the half dozen that failed to meet demands for capacity in the Northeast (see Figures 4.7 to 4.10 for Transco alone, and the earlier figures for spikes at the same time on Tetco and Texas Eastern).

Comparable gains were incurred by the shippers of gas if they contracted firm space previously at the electronic boards at the market hubs. Their transactions, allowing them to sell $[Q_{t-1} + \Delta Q_t]$ for ΔP more on spot gas, did not cause any comparable increases in space charges when subject to price caps on firm contracts. With more than three quarters of the gas at peak under firm contracts for transportation, those shippers gained two to four times the \$6 billion in consumers surplus lost per month from unavailable space on the half dozen large pipelines to the East Coast from Texas.

Then what of those that paid the spike price for the gas of \$8.00 per mcf in Transco Zone Six that month? Because of the lack of capacity, they paid an excess of over the full capacity price of space plus the equilibrium field market spot price. That premium, for all gas sold on the exit hub spot market, has been rolled into quarterly or annual retail charges, which by the time it reaches the consumer's burner tip is in the average merchant service price of the local distributor. The dollar amounts are large, equal to the difference between spike and non-spike spot prices averaged in the volume delivered on an annual basis, roughly twice the consumers surplus foregone. Thus, buyer losses are matched by the rents or excess profits realized by the brokers, dealers, and gas producers who have contracts for space, and ship these volumes at costs capped by FERC rules on firm space contracts. They can purchase or produce at the wellhead, at $2.50 per Mcf, ship it at $1.00 per Mcf, then sell at spot at $8.00 per Mcf, on one or more Bcf shipped per day for as long as the spike lasts.

These are the consequences of the particular process of partial deregulation in the 1990s for the observed performance of the pipelines. The restructuring that took place ended merchant services, and established transactions for gas in spot markets at entry and exit hubs of the pipeline networks. When demand for space increased beyond capacity in these networks, then pipeline charges increased to the cap levels set by FERC, gas was shipped to the entry spot price plus space cap price, but resold at the line exit hub at the market clearing level orders of magnitude greater than entry spot plus cap. The difference was a scarcity rent to the broker-dealer in spot gas. If it had accrued to the pipeline instead, in congestion charges added to firm space contracts without caps (i.e., fully deregulated price), then the additional cash flow would finance or at least justify pipeline expansion. There would then have been no basis differential price spikes after line expansion.

CHAPTER 5

Quantitative Study Number Four

PROFITABILITY OF NATURAL GAS STORAGE RESULTING FROM FEDERAL DEREGULATION IN 1992

*Nickolay Moshkin**

The natural gas storage industry has undergone major deregulatory changes in the last decade, initiated by Order 636 *et al.*, issued by the Federal Energy Regulatory Commission (FERC) in 1993. The objective of this chapter is to empirically examine changes in the natural gas storage operations caused by these regulatory changes. We hypothesize that open access to storage facilities put pressure

*Nickolay Moshkin is a Senior Manager at Cornerstone Research, New York. The material discussed herein may not reflect the opinions of Cornerstone Research.

on the profitability of the incumbent storage sector of the industry. Owners of storage facilities, either the interstate pipelines or retail gas service suppliers, had to face an open market for space. The storage segment of the natural gas industry evolved from vertically integrated segments of pipelines to independent enterprises that had to respond separately to demand fluctuations in consuming gas markets.

The incumbent owners of storage facilities located in consuming states responded to new sources of service in the 1994–1999 period. The evidentiary effects of incumbent response are found in the gas spot price difference between high and low demand periods. If this difference declined, indicating increases in the supply and reductions in prices among storage providers, the deregulated initiative had the intended effect.

To the contrary spot gas prices in producing regions declined in low demand (summer) periods in the years following Order 636 open access. Taken out of historical context, and combined with unchanged winter prices, this fact alone would point to an increase in storage prices, and thus decreased supply of service in the storage markets located in producing states.

Yet, we believe that such a conclusion is incorrect, if the historical context is taken into account. An alternative explanation is that the change in season-to-season price differences has been a result of forecast high and rising oil prices which is expected to cause increases in natural gas demand growth rates. But projected oil prices (up to $50 per barrel) were never realized and industrial and commercial substitution of gas for oil fell far short of the level expected. As a result, the gas industry experienced considerable supply surplus at the wellhead, which has been commonly referred to as "the natural gas bubble." The first stage of "the bubble" primarily affected prices in the low-demand period. In turn, decreased prices in producing regions during summer months combined with access to new storage facilities located there encouraged new entrants marketers to expand producing area storage operations to take advantage of the low price levels. Our data indicate that storage use in producing states increased in this period by about 10–12 percent. The proportion of storage facilities that lost money from storage operations increased from 44.3 percent in 1991–1993 to 65.0 percent in 1994–1999, providing a further sign of intensified supply of storage at the wellhead.

The remainder of the chapter presents theoretical models of storage, followed by the empirical tests for the applications of these models to explaining these counter intuitive prices in this post Order 636 period. The last section summarizes the findings.

5.1. Models

Here, we present five models of gas supply to a market with fluctuations in seasonal demand. The common elements of these models are described here, and specific aspects of each are described later within corresponding subsections. There are two periods, and demand is anticipated to be larger in the second period (inspired by much higher heating rates). In the second period, the demand increase is anticipated to be so significant that even full capacity operations of the pipelines are not able to satisfy it, and therefore storage emerges naturally as a method of inter temporal supply.

The five models differ in ownership patterns and competitive regimes. As such, they provide insight into observed price and quantity behavior under various regimes. The relevant feature of FERC Order 636 was a requirement to separate ownership of gas production from interstate pipeline facilities. The order also required the interstate pipelines that had been the primary owners and operators of storage to dispose of their storage or to rent their reservoir space in the open market to any interested parties. One of the objectives of this study is to analyze the implied effects of the order on prices and quantities supplied of this storage.

5.1.1. *Model 1: Single supplier of transportation owns all storage*

In the first model, we assume that a single pipeline provides gas supply (i.e., product and transport) and storage. Let demand for these services, termed "merchant" services, be linear, deterministic, with the inverse demand function given by:

$$P_D = b_i - aQ_D, \tag{5.1}$$

where $b_i \in \{\mathrm{b}_L, \mathrm{b}_H\}$, with $\mathrm{b}_L < b_H$, is the vertical intercept of the demand curve that is lower in the first period and higher in the second. Let supply be provided by a pipeline that is constrained by capacity, with the following marginal cost function:

$$\mathrm{MC}(Q_S) = \begin{cases} c, & \forall Qs \leq Q_{\max}, \\ \infty, & \forall Qs > Q_{\max}, \end{cases} \tag{5.2}$$

where c is some positive constant.

Let us first consider the case where the product cannot be stored between the two periods and, therefore, cannot be held off the market in period one for future resale. Under these conditions, the optimization problem of service supplier is as follows:

$$\max_{Q_1, Q_2} (bL - aQ_1)Q_1 + (bH - aQ_2)Q_2 - c(Q_1 + Q_2) \tag{5.3}$$

subject to the following set of constraints:

$$Q_1 \in [0, Q_{\max}] \quad \text{and} \quad Q_2 \in [0, Q_{\max}].$$

Because the objective function is separable in the arguments, the choices of the two quantities are independent of each other. The solution is:

$$Q_1 = \min\left\{Q_{\max}, \frac{b_i - c}{2a}\right\}, \quad \text{with } i \in \{L, H\}. \tag{5.4}$$

The corresponding prices are:

$$P_i = \max\left\{b_i - aQ_{\max}, \frac{b_i + c}{2}\right\}, \tag{5.5}$$

and the profits:

$$p_i i = \begin{cases} Q_{\max}(b_i - aQ_{\max} - c), & \text{if } Q_{\max} \leq \dfrac{b_i - c}{2a}, \\ \dfrac{1}{a}\left(\dfrac{b_i - c}{2}\right)^2, & \text{otherwise.} \end{cases} \tag{5.6}$$

Note that the second entry in each formula represents simply the optimal choice by a single pipeline unconstrained by quantities.

Let $Q_{\max}$ now be insufficient to satisfy the second period (high) demand, and allow storage to be available at no cost to the pipeline. It has three control variables,

Q_1, Q_2, and ΔQ, and the objective function is:

$$\max_{Q_1, Q_2, \Delta Q} (b_L - aQ_1)Q_1 + (b_H - a(Q_2 + \Delta Q))$$
$$(Q_2 + \Delta Q) - c(Q_1 + Q_2 + \Delta Q) \tag{5.7}$$

subject to the following set of constraints:

$$Q_1 \geq 0, \Delta Q \geq 0, (Q_1 + \Delta Q) \leq Q_{\max} \quad \text{and}$$
$$Q_2 \in [0, Q_{\max}].$$

When at the optimal solution the constraint $(Q_1 + \Delta Q) \leq Q_{\max}$ is not binding, storage resolves the restriction in supply in the second period effectively, without influencing pricing in the low demand period. The prices observed would be those of an unconstrained supplier.

Otherwise, if there is a limit on supply in the first period to satisfy the current demand and storage requests, the low demand period equilibrium price will be higher than that price, and the high demand period equilibrium price is higher than the unconstrained high demand price as well.

To sum up these observations,

$$P_1^M \leq P_1^{NS} \leq P_1^S \text{ and } P_2^M \leq P_2^S \leq P_2^{NS},$$
$$\text{with } P_1^S \leq P_2^S, \tag{5.8}$$

where the subscripts stand for the period, and the superscripts stand for: *M* – supplier is a single source, with no capacity constraint; *NS* – no storage is available; and

Figure 5.1. Supply is Capacity Constrained ($2Q_{\max}<Q_1^M+Q_2^M$)

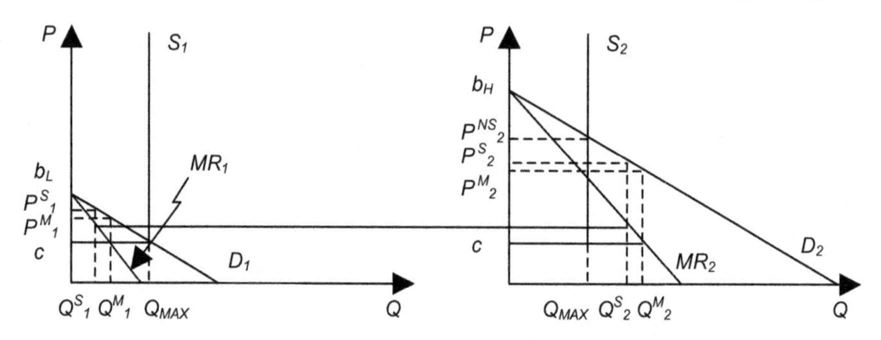

S – storage of the product is possible (see Figure 5.1 for graphical representation of the model).

5.1.2. *Model 2: Single transportation service supplier and single source of storage are operated by two separate owners*

In this model, supply and storage are provided by two independent, service suppliers. Not surprisingly, if the "first mover" is the gas and transport supplier, then it has a significant advantage. Let us show how it works in our framework and what the implications are.

We make the following assumptions about the timing of the events. First, the market service supplier provides quantity Q_1 on the market in the first period. Observing that quantity, the owner of the storage claims any quantity ΔQ_1, up to the total supplied quantity Q_1, to buy and store for future resale. In the second period, the storage owner supplies ΔQ_2, up to the available quantity ΔQ_1, in addition to the quantity Q_2 chosen by the supplier.

The objective function of the merchant supplier has two control variables Q_1 and Q_2:[1]

$$\begin{aligned}\max_{Q_1, Q_2} &\{b_L - a[Q_1 - \Delta Q_1(Q_1)]\} Q_1 \\ &+ \{b_H - a[Q_2 + \Delta Q_2(\Delta Q_1)]\} Q_2 \\ &-c(Q_1 + Q_2) \end{aligned} \tag{5.9}$$

subject to the following set of constraints:

$$Q_1 \in [0, Q_{\max}],\ Q_2 \in [0, Q_{\max}],\ \Delta Q_1 \in [0, Q_1],\ \text{and}\ \Delta Q_2 \in [0, Q_1].$$

Let us solve the problem of the optimal values of Q_1 and Q_2 by backward induction. The objective function of the storage owner in the second period is:

$$\max_{\Delta Q_2} [b_H - a(Q_2 + \Delta Q_2)]\, \Delta Q_2 \tag{5.10}$$

subject to the following constraint:

$$\Delta Q_2 \in [0, Q_1].$$

---

[1] We do not explicitly assume here that in the second period the storage owner would sell all the gas purchased in the first, namely, $\Delta Q_1 = \Delta Q_2 = \Delta Q$. Ultimately, that is what happens.

The objective function of the merchant supplier in the second period is:

$$\max_{Q_2}[b_H - a(Q_2 + \Delta Q_2) - c]Q_2 \tag{5.11}$$

subject to the following constraint:

$$\Delta Q_2 \in [0, Q_{\max}].$$

The solution for the quantities is:[2]

$$\Delta Q_2 = \min\left\{\frac{b_H + c}{3a}, \Delta Q_1\right\}, \tag{5.12}$$

$$Q_2 = \max\left\{\frac{b_H - 2c}{3a}, \frac{b_H - a\Delta Q_1 - c}{2a}\right\}, \tag{5.13}$$

and for the prices:

$$P_2 = \max\left\{\frac{b_H + c}{3}, \frac{b_H - a\Delta Q_1 + c}{2}\right\}. \tag{5.14}$$

The profits of the two players are:

$$\pi_{\Delta 2}(\Delta Q_1) = \begin{cases} \frac{1}{a}\left(\frac{b_H + c}{3}\right)^2, & \text{if } \frac{b_H + c}{3a} \leq \Delta Q_1, \\ \left(\frac{b_H - a\Delta Q_1 + c}{2}\right)\Delta Q_1, & \text{otherwise}, \end{cases} \tag{5.15}$$

[2]We derive the solution for the case when $Q_{\max}$ is not binding.

and

$$\pi_2 = \begin{cases} \dfrac{1}{a}\left(\dfrac{b_H - 2c}{3}\right)^2, & \text{if } \dfrac{b_H + c}{3a} \leq \Delta Q_1, \\ \dfrac{1}{a}\left(\dfrac{b_H - a\Delta Q_1 - c}{2}\right)^2, & \text{otherwise.} \end{cases} \tag{5.16}$$

Interpretation of these results is as follows. The first case represents the simultaneous duopoly game, with players having a constant marginal cost of production equal to c and 0, up to capacities $Q_{\max}$ and ΔQ_1, respectively. Equilibrium is achieved at levels of production under corresponding firms' capacities. The second case is similar, only parameter values are chosen so that the player with zero marginal cost 'produces' at full capacity, ΔQ_1. The second firm then behaves as a monopoly facing the demand curve with a vertical intercept at $b_H - a\Delta Q_1$.

Next, we proceed to find the optimal storage quantity of the storage owner in the first period. The objective function is:

$$\max_{\Delta Q_1} \pi_{\Delta 2}(\Delta Q_1) - (b_L - a(Q_1 - \Delta Q_1))\Delta Q_1. \tag{5.17}$$

This form shows that acquiring more than $\frac{b_H+c}{3a}$ of gas in the first period cannot be optimal: there is no use for the extra gas, the storage owner has to pay a higher price for it. Therefore, we plug in the second value for $\pi_{\Delta 2}(\Delta Q_1)$

in the objective function as follows:[3]

$$\max_{\Delta Q_1} \left\{ \left[\frac{(b_H - a\Delta Q_1 + c)}{2} \right] - [b_L - a(Q_1 - \Delta Q_1)] \right\} \Delta Q_1. \tag{5.18}$$

Taking the F.O.C., we find:

$$\left(\frac{b_H - 2a\Delta Q_1 + c}{2} \right) - [b_L - a(Q_1 - 2\Delta Q_1)] = 0, \tag{5.19}$$

or

$$\Delta Q_1(Q_1) = \frac{b_H - 2b_L + c}{6a} + \frac{Q_1}{3}. \tag{5.20}$$

The optimal amount of storage ΔQ_1 increases with the gas quantity provided in the market by the supplier in the first period; it increases with the vertical intercept of the second period, and decreases with the vertical intercept of the first period. The explanation is that the larger first period supply of the good Q_1 implies lower purchasing price for the storage owner. Also higher b_H implies higher selling price in the second period, and lower b_L implies lower purchase prices for the suppliers in the first period, and therefore higher profits.

---

[3] After we find the solution, we will verify the indeed $\frac{b_H + c}{3a} > \Delta Q_1$.

Having all the best responses at hand, the merchant service supplier chooses the quantity for the first period:

$$\max_{Q_1} \{b_L - a[Q_1 - \Delta Q_1(Q_1)] - c\} Q_1 + \frac{1}{a}\left[\frac{b_H - a\Delta Q_1(Q_1) - c}{2}\right]^2. \tag{5.21}$$

Taking the F.O.C., we arrive at the equation:

$$\{b_L - a[Q_1 - \Delta Q_1(Q_1]) - c\} - \frac{2}{3}aQ_1 + \frac{2}{a}\left[\frac{b_H - a\Delta Q_1(Q_1) - c}{2}\right]\left(\frac{-a}{6}\right) = 0. \tag{5.22}$$

Simplifying, we get:

$$36b_L - 36c + 6b_H - 12b_L + 6c - 6b_H + b_H - 2b_L + c + 6c = 46aQ_1 \tag{5.23}$$

or

$$Q_1 = \frac{b_H + 22b_L - 23c}{46a} = \frac{b_H - b_L}{46a} + \frac{b_L - c}{2a} > Q_1^M = \frac{b_L - c}{2a}. \tag{5.24}$$

Thus, the optimal choice of quantity supplied in the first period exceeds that for the case of no possibility of storage.

Returning to other variables, we find that:

$$\Delta Q_1(Q_1) = \frac{b_H - 2b_L + c}{46a} + \frac{1}{3}\frac{b_H - b_L}{46a} + \frac{b_L - c}{6a}$$
$$= \frac{24}{23a}\frac{1}{6}(b_H - b_L)$$
$$= \frac{4}{23a}(b_H - b_L) > 0. \tag{5.25}$$

This confirms that the quantity stored is below the unconstrained optimal supply in the second period, $(b_H + c)/3a$, making our choice of second period profit for the storage owner in the optimization problem valid.

The quantity provided by the supplier in the second period is:

$$Q_2 = \frac{b_H - a\Delta Q_1 - c}{2a}$$
$$= \frac{1}{2a}\frac{19b_H + 4b_L - 23c}{23} < Q_2^M$$
$$= \frac{1}{2a}(b_H - c). \tag{5.26}$$

The equilibrium prices of the first period rises above the no storage prices:

$$P_1 = \frac{1}{46}(7b_H + 16b_L + 23c) > P_1^M = \frac{1}{2}(b_L + c), \tag{5.27}$$

while the price of the second period falls below it:

$$P_2 = \frac{1}{46}(19b_H + 4b_L + 23c) > P_2^M = \frac{1}{2}(b_L + c). \tag{5.28}$$

The difference between high and low spot gas prices in the separate ownership model is smaller than in joint ownership model. We would like to test this hypothesis on real data, namely, that the separation of ownership established by Order 636 would have resulted in lower observed price differences between winter and summer prices.

5.1.3. *Model 3: Storage in the presence of competitive supply of gas plus transportation*

In this case, storage activity is at all necessary only when $Q_{\max}$ falls short of satisfying the market demand in the high demand period:

$$c < b_H - aQ_{\max}, \tag{5.29}$$

or

$$Q_{\max} < \frac{b_H - c}{a}. \tag{5.30}$$

The storage owner would choose ΔQ to maximize the following objective function:

$$\max_{\Delta Q}(\max\{[b_H - a(Q_{\max} + \Delta Q)], c\} \\ - \max\{[b_L - a(Q_{\max} - \Delta Q)], c\})\Delta Q. \tag{5.31}$$

Let us assume that the second period price exceeds c at equilibrium, for otherwise the storage owner makes no profit, and the problem degenerates into one with equilibrium prices $P_1 = P_2 = c$. Let us also assume that the first period price exceeds c. The solution, under these conditions, is found from the F.O.C.:

$$[b_H - a(Q_{\max} + 2\Delta Q)] - [b_L - a(Q_{\max} + 2\Delta Q)] = 0, \tag{5.32}$$

or

$$\Delta Q = \frac{b_H - b_L}{4a}. \tag{5.33}$$

The prices will be

$$P_1 = \frac{b_H + 3b_L}{4} - aQ_{\max} > P_1^{NS} = b_L - aQ_{\max}, \tag{5.34}$$

and

$$P_2 = \frac{3b_H + b_L}{4} - aQ_{\max} < P_2^{NS} = b_H - aQ_{\max}. \tag{5.35}$$

When the first period supply is sufficient to satisfy both current demand and storage requests,

$$Q_{\max} > \frac{b_L - c}{a} + \Delta Q, \tag{5.36}$$

the price in the first period is simply c. The F.O.C.s are:

$$[b_H - a(Q_{\max} + 2\Delta Q)] - c = 0 \tag{5.37}$$

with

$$\Delta Q = \frac{b_H - aQ_{\max} - c}{2a}, \tag{5.38}$$

which is equal to the supply of a single merchant source of gas plus transportation with marginal cost of production c

Figure 5.2. Net Increase in Consumer Surplus due to Storage

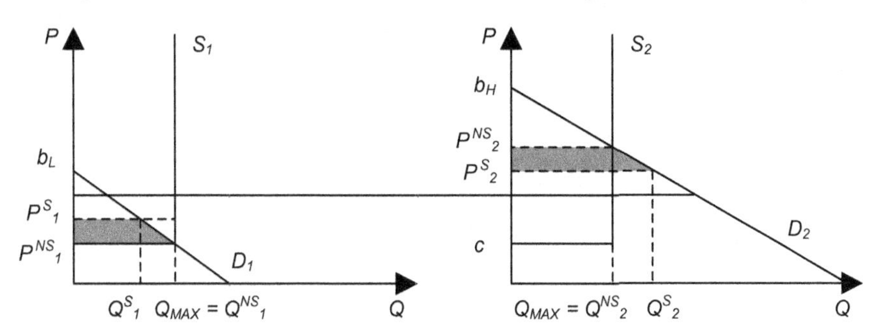

into the market with vertical intercept at $b_H - aQ_{\max}$. The second period price thus is:

$$P_2 = \frac{b_H - aQ_{\max} + c}{2} < P_2^{NS} = b_H - aQ_{\max}. \quad (5.39)$$

The presence of storage causes period prices to converge in comparison with the no storage case, which increases consumer surplus. Figure 5.2 shows the representative picture corresponding to Model 3; analysis of the other models provides similar results.

5.1.4. *Model 4: Single merchant service supply in the presence of competitive storage*

Competitive storage equalizes the prices in two periods, for otherwise entry into the storage market would continue to occur and storage capacity would expand. Numerous storage operators make zero profits at equilibrium. The objective function of the supplier is:

$$\begin{aligned}\max_{Q_1,Q_2} & \{b_L - a[Q_1 - \Delta Q(Q_1, Q_2)]\}Q_1 \\ & +\{b_H - a(Q_2 + \Delta Q(Q_1, Q_2)]\}Q_2 \\ & -c(Q_1 + Q_2). \end{aligned} \tag{5.40}$$

As we mentioned above, the ΔQ can be found from the equality of prices in the two periods:

$$b_L - a(Q_1 - \Delta Q) = b_H - a(Q_2 - \Delta Q), \tag{5.41}$$

or

$$\Delta Q(Q_1, Q_2) = \frac{b_H - b_L}{2a} - \frac{Q_1 - Q_2}{2}. \tag{5.42}$$

Plugging this into the objective function, we find:

$$\begin{aligned}\max_{Q_1,Q_2} & (b_L - a(Q_1 - \Delta Q))Q_1 \\ & + (b_H - a(Q_2 + \Delta Q))Q_2 - c(Q_1 + Q_2). \end{aligned} \tag{5.43}$$

The F.O.C.s are:

$$b_L - aQ_1 + a\Delta Q - \frac{a}{2}Q_1 - \frac{a}{2}Q_2 - c = 0, \tag{5.44}$$

and

$$-\frac{a}{2}Q_1 + b_H - aQ_2 + a\Delta Q - \frac{a}{2}Q_2 - c = 0. \tag{5.45}$$

these two equations degenerate into one:

$$Q_1 + Q_2 = \frac{b_H + b_L - 2c}{2a}, \tag{5.46}$$

with the price in two periods being:

$$P = \frac{b_H + b_L - 2c}{4}. \tag{5.47}$$

Therefore, in this model the prices should be equal in the high and low demand period. Consumer surplus increases compared to the no storage case. This should come as no surprise, for at equilibrium the prices in two periods are the same, and the supplier is indifferent in which period to supply the product. The competitive storage will readjust any occurring disbalances. Thus, the only variable that matters for the maximization problem is the total supply in both periods, but not the individual period supply.

5.1.5. *Model 5: Perfectly competitive merchant supply and storage*

In this model, both transportation and storage activities are assumed to be perfectly competitive, providing no profits for all the parties involved. The prices in the two periods will be equalized, just as in the previous model. However, the magnitude of the prices in Model 5 is lower. Let us assume that $Q_{\max}$ is relatively small, so that in both periods suppliers will provide that amount to the market. Prices are equal and may be found from the equation for the storage quantity:

$$b_L - a(Q_{\max} - \Delta Q) = b_H - a(Q_{\max} + \Delta Q). \quad (5.48)$$

It immediately follows that:

$$\Delta Q = \frac{b_H - b_L}{2a}, \quad (5.49)$$

and

$$P = \max\left\{\frac{b_H - b_L}{2} - aQ_{\max}, c\right\}. \quad (5.50)$$

Hence, in the model again the prices in the high and low demand periods are equalized.

5.1.6. *Comparison of the various models*

Table 5.1 summarizes the anticipated prices and high/low demand prices differences for the various models. It directly follows from these models that the spread in prices of high and low demand is determined mainly by the competitive regime in the storage market. The storage owner is pivotal in deciding the degree of the inter-period product substitution. The price difference turns out to be the highest in cases when storage is operated by a single supplier facing a number of highly competitive merchant service suppliers. The price differential is reduced by almost half when the storage owner faces a single merchant service supplier in the market. The cases of oligopolistic (Cournot) supply would fall in

Table 5.1. Equilibrium Prices

Model	**Prices**		
	Low Demand	**High Demand**	**Difference**
Integrated Supplier/ Storage	$1/2(b_L + c)$	$1/2(b_H + c)$	$1/2(b_H - b_L)$
Separate Supplier/Storage	$1/46(7b_H + 16b_L + 23c)$	$1/46(19b_H + 4b_L + 23c)$	$6/23(b_H - b_L)$
Monopoly + Competitive	$1/4(b_H + b_L + 2c)$	$1/4(b_H + b_L + 2c)$	0
Competitive + Monopoly	$1/4(b_H + 3b_L) - aQ_{\max}$	$1/4(3b_H + b_L) - aQ_{\max}$	$1/2(b_H - b_L)$
Competitive + Competitive	$1/2(b_H + b_L) - aQ_{\max}$	$1/2(b_H + b_L) - aQ_{\max}$	0

between the two. Competitive storage, by the nature of perfect competition, equalizes prices in the two periods.

Relative prices also can be analyzed on the basis of these models. The price differentials in the jointly owned single merchant service supplier with own storage lie outside the interval between prices in the separate firms case, and furthermore, the prices in the single merchant supply with competitive storage case lie inside both intervals. The prices in the case of single supply with competitive storage exceed the prices with competitive markets.

Thus, we have a set of testable hypotheses that will indicate the impacts of Order 636 both in supply/transportation and storage markets. Small or zero differences between high and low spot gas prices would indicate that the storage market is competitive, and that the deregulatory policy of "open access" has been successful in the storage market. The magnitude of the spread provides an assessment of the combined deregulatory effects in these two segments of the markets. The time trends of prices for each of the low and high demand seasons clarify the effects further. An increase in the price of gas in the low demand period, or decrease in the high demand period, is consistent with a deregulatory determined increase in competition in the storage market. The reverse movement points toward an increase in the market power of corresponding storage entities contrary to the intent of the order.

5.2. Empirical Tests

In this section, we report on empirical results and test their consistency with the predictions from the models. Such results should identify changes in the competitive regimes induced by FERC Order 636. The 1991–1999 heating year[4] data are used for the purposes of this exercise.

The data set consists of monthly prices and volumes in gas storage for 19 states with major storage reservoirs.[5] The storage volumes are defined as the difference between the total storage capacity and base gas requirements for performing the storage function. The states with storage capacities exceeding 40,000 MMcf are included in the study. The set is further divided into four subgroups by the total size of storage in a state (large or medium) and by state affiliation (gas producing or gas consuming),[6] with division by size at 100,000 MMcf. The states included in various groups are listed in Table 5.2.

---

[4]The heating year defined in this work starts in April.

[5]*Source*: Taken originally from forms EIA-857 and EIA-895, "*Monthly Report of Natural Gas Purchases and Deliveries to Consumers*", and reported in various issues of EIA publication "*Natural Gas Monthly*".

[6]A state is defined as "producing" if a quantity of gas produced in the state exceeds gas consumption of the state.

Table 5.2. Groups

Group	**States Included**
Producing, Large Storage	WV, TX, LA, OK, KS
Producing, Medium Storage	NM, UT, CO, WY
Consuming, Large Storage	MI, PA, IL, OH, CA, KY
Consuming, Medium Storage	NY, IA, MS, IN

Using these data, we generate the weighted average high demand ("winter") and low demand ("summer") prices, with corresponding periods defined from November to the following March and from April to October. The volume weights used are injections to or withdrawals from the storage facilities in a state; they approximate the per unit revenues of storage enterprises. The storage usage has been measured by the variation in the intake and outtake of the gas defined as a sum of the absolute values for injections to and withdrawals from a storage facility in a given heating season.

5.2.1. *Change in differences between winter and summer gas prices*

The dependent variable in the following regressions is the difference between gas spot prices in high and low demand periods, and the independent variable is "change in regulation". Throughout the analysis, this is a binary variable, with a value of zero for the years before Order 636 (1991 through 1993), and a value of one for all consecutive years thereafter (1994 through 1999).

Considering first the consuming regions, the results of the regressions have been collected in Table 5.3. They show

Table 5.3. Price Differences Regressions, Consuming States

Variable	**P_H–P_L, \$/Mcf**					
	All		**Large**		**Medium**	
	Coefficient	**Std. Error**	**Coefficient**	**Std. Error**	**Coefficient**	**Std. Error**
MI	0.295	0.139	0.325	0.138		
PA	−0.807	0.139	−0.777	0.138		
IL	0.043	0.139	0.073	0.138		
OH	−0.483	0.139	−0.454	0.138		
CA	0.137	0.139	0.167	0.138		
KY	0.089	0.139	0.118	0.138		
NY	0.072	0.139			0.027	0.171
IA	−0.737	0.139			−0.782	0.171
MS	0.0174	0.139			0.129	0.171
IN	−0.026	0.139			−0.071	0.171
$D_{1994-1999}$	−0.026	0.085	−0.072	0.104	0.040	0.148
# of Obs.	90		54		36	
$\bar{R}^2$, %	46.7		52.4		39.1	

that in the post Order 636 period of open access to gas storage and transportation the price spread between winter and summer gas prices declined on average by 2.6 cents per thousand cubic feet (Mcf) for all consuming regions, declined by 7.2 cents per Mcf for a sub-sample of consuming states with large storage facilities, but increased by about 4 cents per Mcf for consuming states with medium sized storage facilities. These changes are consistent with a shift of the storage market towards more capacity and more competition in the large consuming states. The regression equations explain about 40 to 50 percent of the variance in the dependent variable, but the coefficients on the regulatory variable are not statistically significant.

Further, there is an indication that storage facilities in Pennsylvania, Ohio, and Iowa may have lost money from their post-Order 636 operations. For instance, it follows from the negative coefficient on the states variable in the regression on gas stored and resold in the 1991–1993 period. There are plausible explanations for such reductions (other than poor management); for one, local distribution companies (LDCs) are pipelines that have firm contractual obligations to satisfy the demand of large numbers of residential and industrial customers, and do so at any price to avoid extensive penalties for the failure to meet these obligations.

The regression results for the producing states are summarized in Table 5.4. We find that the price spread for the producing states has increased on average by 64.3 cents per Mcf for the complete sample of producing regions, by 35.2 cents per Mcf for a sub-sample of producing states with large storage facilities, and by about 1 dollar per Mcf for producing states with medium storage facilities.

Table 5.4. Price Differences Regressions, Producing States

Variable	**P_H–P_L, \$/Mcf**					
	All		**Large**		**Medium**	
	Coefficient	**Std. Error**	**Coefficient**	**Std. Error**	**Coefficient**	**Std. Error**
WV	−0.935	0.267	−0.741	0.247		
TX	−0.034	0.267	0.160	0.247		
LA	−0.232	0.267	−0.038	0.247		
OK	−0.044	0.267	0.150	0.247		
KS	−0.482	0.267	−0.288	0.247		
NM	−0.348	0.267			−0.592	0.327
UT	−1.040	0.267			−1.284	0.327
CO	−0.577	0.267			−0.821	0.327
WY	−0.257	0.267			−0.501	0.327
$D_{1994-1999}$	0.643	0.175	0.352	0.198	1.009	0.284
# of Obs.	81		45		36	
$\bar{R}^2$, %	22.3		20.2		28.3	

Coefficients on the regulatory dummy in all three regressions are not only positive but also statistically significant. This result implies that storage after the Order 636 regulatory change in producing states became controlled by few sources of supply of service. The regressions explain between 20 and 30 percent of the variance in the dependent variable.

These regressions also show that the usage of storage facilities located in the producing regions has become more profitable in the period after Open Access. That, in turn, should induce more extensive storage. We test this hypothesis later on the data on storage usage.

5.2.2. *Changes in winter prices*

The results of regressing winter prices in consuming states on state and regulation dummies are presented in Table 5.5. Winter prices in consuming regions have declined, but the estimated decline is not statistically significant. For the total sample of consuming regions the point estimate of the decline is 6.5 cents per Mcf; for the sub-sample with large storage the estimate is 3.8 cents per Mcf, and for the sub-sample with medium storage facilities it is 10.7 cents per Mcf. The relative price decline is about 5 percent of the average winter prices. The gas prices range from $2.60 per Mcf in California to $3.86 per Mcf in Ohio. The first two

Table 5.5. Winter Price Regressions, Consuming States

Variable	**P_H, \$/Mcf**					
	All		**Large**		**Medium**	
	Coefficient	**Std. Error**	**Coefficient**	**Std. Error**	**Coefficient**	**Std. Error**
MI	2.92	0.14	2.90	0.15		
PA	3.24	0.14	3.22	0.15		
IL	2.90	0.14	2.89	0.15		
OH	3.86	0.14	3.84	0.15		
CA	2.60	0.14	2.58	0.15		
KY	3.13	0.14	3.11	0.15		
NY	2.91	0.14			2.94	0.15
IA	3.03	0.14			3.06	0.15
MS	2.83	0.14			2.86	0.15
IN	2.77	0.14			2.80	0.15
$D_{1994-1999}$	−0.065	0.086	−0.038	0.115	−0.107	0.129
# of Obs.	90		54		36	
$\bar{R}^2$, %	38.5		46.2		−2.2	

regressions explain about 40 to 45 percent of the variance, while the last regression has very little explanatory power, with a negative (adjusted) R-squared.

The results of price regressions for winter producing states are summarized in Table 5.6. Winter prices in the sample of all producing regions have declined by about 5.4 cents per Mcf, increased by 2.9 cents per Mcf in the large

Table 5.6. Winter Price Regressions, Producing States

Variable	P_H, \$/Mcf					
	All		Large		Medium	
	Coefficient	Std. Error	Coefficient	Std. Error	Coefficient	Std. Error
WV	3.16	0.15	3.11	0.16		
TX	3.14	0.15	3.08	0.16		
LA	2.65	0.15	2.59	0.16		
OK	2.58	0.15	2.52	0.16		
KS	2.75	0.15	2.70	0.16		
NM	2.11	0.15			2.18	0.17
UT	2.97	0.15			3.04	0.17
CO	2.64	0.15			2.71	0.17
WY	2.99	0.15			3.06	0.17
$D_{1994-1999}$	−0.054	0.096	0.029	0.127	−0.158	0.146
# of Obs.	81		45		36	
$\bar{R}^2$, %	33.0		21.6		40.5	

storage fields sample, and declined by 15.8 cents per Mcf for the sample of producing states with medium-sized storage. However, all coefficients are statistically insignificant. The regressions explain from 21.6 to 40.5 percent of the variance of the dependent variable.

These regressions establish no more than a general tendency for decline in winter price levels across producing and consuming states, and storage field sizes. The price declines are small relative to price variations, and therefore, no statistically significant inferences can be made at this stage of the analysis.

5.2.3. *Changes in summer prices*

The results of summer price regressions for consuming states are summarized in Table 5.7. Summer prices in con-

Table 5.7. Summer Price Regressions, Consuming States

Variable	**P_L, \$/Mcf**					
	All		**Large**		**Medium**	
	Coefficient	**Std. Error**	**Coefficient**	**Std. Error**	**Coefficient**	**Std. Error**
MI	2.62	0.17	2.58	0.17		
PA	4.04	0.17	4.00	0.17		
IL	2.86	0.17	2.81	0.17		
OH	4.34	0.17	4.29	0.17		
CA	2.46	0.17	2.42	0.17		
KY	3.04	0.17	2.99	0.17		
NY	2.84	0.17			2.91	0.19
IA	3.77	0.17			3.84	0.19
MS	2.66	0.17			2.73	0.19
IN	2.80	0.17			2.87	0.19
$D_{1994-1999}$	−0.038	0.103	0.034	0.131	−0.147	0.166
# of Obs.	90		54		36	
$\bar{R}^2$, %	63.7		70.4		44.7	

suming regions have not changed significantly, with the point estimate for the sample of all consuming states at 3.8 cents per Mcf decline in the deregulation period, 3.4 cents increase for the sub-sample of consuming states with large storage fields, and 14.7 cents decline for the sub-sample of consuming states with medium-sized storage. Forty-five to 70 percent of the variance is explained by the regressors.

Table 5.8 reports the results of summer price regressions for producing states. Summer prices in producing regions have changed quite a bit. This is the first set of regressions in the analysis of winter or summer price changes that shows clear statistical evidence of a decrease in the dependent variable in the second, deregulatory period. For all three samples, the coefficients on the regulatory dummy are highly statistically significant. The summer price decline in the deregulation period for the sample of all producing states is estimated at 69.8 cents per Mcf, 32.3 cents per Mcf for the sub-sample of producing states with large storage fields, and 116.6 cents for the sub-sample of states with medium-sized storage. About 42 to 52 percent of the variance is explained by the regressors.

Such a change of prices in producing regions is indicative of results from the earlier deregulation of wellhead prices in NGPA (1978) (Figures 1.1 and 1.2 in MacAvoy [2000]). Our theoretical models are not immediately applicable in

Table 5.8. Summer Price Regressions, Producing States

	P_L, \$/Mcf					
Variable	**All**		**Large**		**Medium**	
	Coefficient	**Std. Error**	**Coefficient**	**Std. Error**	**Coefficient**	**Std. Error**
WV	4.10	0.25	3.85	0.19		
TX	3.17	0.25	2.92	0.19		
LA	2.88	0.25	2.63	0.19		
OK	2.62	0.25	2.37	0.19		
KS	3.24	0.25	2.99	0.19		
NM	2.46	0.25			2.77	0.33
UT	4.01	0.25			4.32	0.33
CO	3.22	0.25			3.53	0.33
WY	3.24	0.25			3.56	0.33
$D_{1994-1999}$	−0.698	0.160	−0.323	0.151	−1.166	0.287
# of Obs.	81		45		36	
$\bar{R}^2$, %	42.1		52.4		45.2	

this framework, for we consider a two-period static model, while the real scenarios incorporate dynamic prices in this period. However, the intuition obtained here will be helpful even in a more complicated world.

Thus, data show that the prices at the wellhead in the summer periods went down significantly so that participants in the gas spot markets should have found it beneficial to get more involved in storage demands. If this was the case, it was profitable to purchase gas at the wellhead in the summer period, transport it to the point of consumption, and store it in consuming regions until use in the winter. Another likely alternative would have been to purchase gas at the wellhead in the summer, store it in producing regions' reservoirs, then transport it to the consuming regions in the winter period for immediate consumption (subject to pipeline winter capacity constraints). What market conditions would influence the course of actions by the firms operating in the market? For one, relative winter/summer transportation prices. For another, storage availability and prices. Let us look into these issues in the next subsection.

5.2.4. *Changes in transportation prices*

Applying the standard assumption of arbitrage in transportation markets, we use as a proxy for the transportation charge the differences between the gas prices at destination and origin. According to the theory, this change should decline if Order 636 was effective in storage markets. In order to discern the direction of the change in the transportation price, we assume that all the producing regions in the data set deliver gas to all the consuming regions. We define as the dependent variable in a regression explaining use of storage the difference between transportation charges in the winter and in the summer.

We call a pair of producing/consuming states a "pipeline" between these locations, and include all pipeline dummies in the regression (a constant is excluded to avoid multicollinearity). The only other explanatory variable included is the regulatory change dummy. The coefficient of this regulation dummy is reported in Table 5.9, while coefficients on individual pipeline dummies are not shown.

The regression indicates that in the post-Order 636 years it became relatively cheaper to transport gas in the winter

Table 5.9. Transportation Price Regressions

TR_H–TR_L, \$/Mcf		
All Interactions		
Variable	**Coefficient**	**Std. Error**
# of Pipeline D's	90	
$D_{1994-1999}$	−0.575	0.051
# of Obs.	810	
$\bar{R}^2$, %	31.1	

than in the summer. It does not imply that transportation prices in the winter went down significantly; in fact, the data show that it was transportation prices in the summer that increased. In any event, the change in transportation costs provided a better economic incentive to purchase gas at the wellhead in the summer, store it in the producing region until the winter, and then transport it for immediate consumption. The final empirical test of this hypothesis is a test that would verify an increase in storage usage in producing regions, and a decrease in such usage for consuming regions. The next subsection describes such a test.

5.2.5. *Changes in storage quantity variations*

One natural variable to approximate "use of storage" is the difference between the maximum and the minimum amounts of gas in storage within the same heating year. In most years, usually the maximum is achieved at the beginning of November, while the minimum occurs at the beginning of April. Here, we again separate producing and consuming regions, and also group states by size of the available storage facilities. The dependant variable is "use", while the independent variable is the regulatory dummy, equal to 1 for the post Order 636 years.

The Durbin-Watson statistic for regressions that include the sample of all consuming states and sub-sample of large states failed the hypothesis of no serial correlation in errors; appropriate adjustments to the estimation procedure have been made, and these results are included in Table 5.10. As follows from these two samples, the storage usage in the consuming states has declined in the 1994–1999 period, on average by 70.9 Bcf per year per state for the whole

Table 5.10. Storage Usage Regressions, Consuming States

Variable	**$MAX_{STORAGE}$–$MIN_{STORAGE}$, Bcf**					
	All		**Large**		**Medium**	
	Coefficient	**Std. Error**	**Coefficient**	**Std. Error**	**Coefficient**	**Std. Error**
MI	816.3	33.8	810.2	29.9		
PA	596.3	33.8	588.3	29.9		
IL	424.3	33.8	426.2	29.9		
OH	390.0	33.8	382.9	29.9		
CA	251.8	33.8	254.5	29.9		
KY	149.6	33.8	144.1	29.9		
NY	155.0	33.8			100.6	3.2
IA	173.8	33.8			113.8	3.2
MS	104.5	33.8			52.6	3.2
IN	88.3	33.8			35.8	3.2
$D_{1994-1999}$	−70.9	21.6	−70.1	23.0	1.34	2.76
# of Obs.	80		48		36	
$\bar{R}^2$, %	88.3		91.1		94.6	

sample, and by 70.1 Bcf per year per state for the sub-sample of states with medium storage. The declines in the first two samples are highly statistically significant. Regressions explain 88 to 95 percent of the variances in the dependent variable.

The results of similar regressions for producing states are summarized in Table 5.11. Here again, we find support

Table 5.11. Storage Usage Regressions, Producing States

Variable	$MAX_{STORAGE} - MIN_{STORAGE}$, Bcf					
	All		Large		Medium	
	Coefficient	Std. Error	Coefficient	Std. Error	Coefficient	Std. Error
WV	261.9	8.9	253.9	12.1		
TX	247.3	8.9	239.2	12.1		
LA	249.9	8.9	241.8	12.1		
OK	155.8	8.9	147.7	12.1		
KS	102.7	8.9	94.7	12.1		
NM	1.4	8.9			11.4	2.7
UT	45.8	8.9			55.9	2.7
CO	34.9	8.9			45.0	2.7
WY	7.7	8.9			17.8	2.7
$D_{1994-1999}$	18.4	5.7	30.5	9.7	3.3	2.3
# of Obs.	81		45		36	
$\bar{R}^2$, %	94.8		81.7		89.0	

of our main hypothesis. The usage of storage increased in all three samples, in the sample of all producing states by 18.4 Bcf per year per state, in the states with large storage field by 30.5 Bcf per year per state, and in the states with medium storage reservoirs by 3.3 Bcf per year per state. These numbers correspond to 10 to 12 percent increase in storage activity. The first two estimates are highly statistically significant, while the last is marginally significant. The included explanatory variables account for 80 to 95 percent of variation in the dependent variable.

5.3. Conclusions

The empirical investigation of the data shows that most of the finding are consistent with the theoretical model presented in the earlier part of the chapter. After the regulatory changes embodied in Orders 636 *et al.* took place, wellhead prices went down, transportation cost went up, and storage in the producing regions in the summer increased. The complicated picture of the "real world", induced by the dynamic nature of the industry, with multiperiod optimization, uncertainty and other factors excluded from the theoretical models, limits the usefulness of these findings.

Yet, the suggested models offer a basis for understanding the empirical results. The cycle of natural gas storage may be approximated reasonably well by the two-period models considered in this paper, given that the process of filling up and emptying storage follows a two season predictable pattern. There is virtually no longer term consideration in determining storage intakes or outtakes — since it is not likely to have been optimal "to save gas this summer for consumption in three years." Therefore, the terminal feature of the two-period model that constrains storage reserves at zero at the end of the second period (end of the winter) is realistic. Due to these specific features of the industry, we believe that the way to simplify the infinite horizon dynamic programming framework is to separate it into a set of disjoint two-period problems, and then solve them.

The models-based interpretation of the empirical findings is as follows. Price differences after Orders 636 et al. between high and low demand periods in consuming states have shown a small negative change, which supports the inference that storage has moved toward lower pricing for service. Price differences between high and low demand periods in producing states have increased; we hypothesize that this change is due to a significant decline in summer wellhead prices, caused by unexpected increased gas field supply of the 1980s.

We also conclude that prices for use of storage facilities located in the producing regions have increased since the regulatory change, probably from increased demand.

The data further indicate that the increased gas supply at the wellhead in the low demand period together with relatively low transportation charges in a high demand period has encouraged firms to switch from storing gas in consuming regions to storing it in producing regions. That conjecture has been supported by the data on storage usage in these territories.

In addition, the direct estimation of profits of the storage facilities shows losses for about 44.3 percent of the storage facilities in 1991–1993, and for about 65.0 percent in 1994–1999. The estimate is a value of

$$\sum_{t \in \text{HEATING\_YEAR}} (P_t \Delta Q_t), \tag{5.51}$$

which is realized profits by a storage facility with (assumed) zero costs of storage. This measure, however, does not include another portion of profits by firms operating storage facilities, which comes from resales of natural gas. Our conjecture is that those profits should dominate the losses calculated in (5.51) from storage. Also, the losses could be an indication that the market faces considerable uncertainty and the presence of unavoidable costs of satisfying fluctuating demand of uninterruptible customers.

CHAPTER 6

Quantitative Study Number Five

REVISING THE MODEL OF GAS WELLHEAD PRICES AND QUANTITIES FOR PARTIAL DEREGULATION*

Paul W. MacAvoy and Vadim Marmer

The characteristic feature of small econometric models for gas that goes into pipelines is that wellhead prices for dedicated reserves bring total production in line with

---

*This chapter has been adapted from the model presented in Chapter 2 of MacAvoy, P. W. (2000). *The Natural Gas Market* (Yale University Press) that presented a traditional full regulatory set of equations for supply and demand. Vadim Marmer developed the new post-regulatory equations.

total retail demand. Certain institutional practices make this process of convergence to equilibrium take place over some period of time. Production is generated under terms of reserves contracts so that supply has to adjust to demands in a multiyear process. Excess demands in times of extreme cold weather are dealt with by increasing take from reserves and by expanding take from storage and from "line pack" in the pipelines under long-term contracts. Excess demands over long periods is eliminated by higher prices in new contracts which trigger price increases in existing contracts; these higher prices not only cause additions to reserves and production but also curtail long-term demands.

FERC Order 636 changed all that. The Order in effect ended contract markets and put in their place spot markets for gas from reserves owned not by pipelines but by producers. This chapter develops post-Order equations and forecasts to account for these changes.

6.1. Estimation of Demand

6.1.1. *Equations*

Consider first this characteristic set of relationships inherent in demands for natural gas. Production demand is marked by separate residential, industrial, commercial, and electric utility sectors. The substitution of gas for other energy sources occurs at different rates in each sector.

To account for these differences, the demand for energy in each sector breaks down into gas and other fuel shares. Share equations for gas, petroleum, coal and electricity differ by sector. Total gas demand equals the sum of these gas volumes over all four sectors.

The equation set constructed to estimate annual demand for natural gas production is shown in Figure 6.1. Four sectoral demand equations depend on "market size", as

Figure 6.1. Flow Diagram: Natural Gas Demand

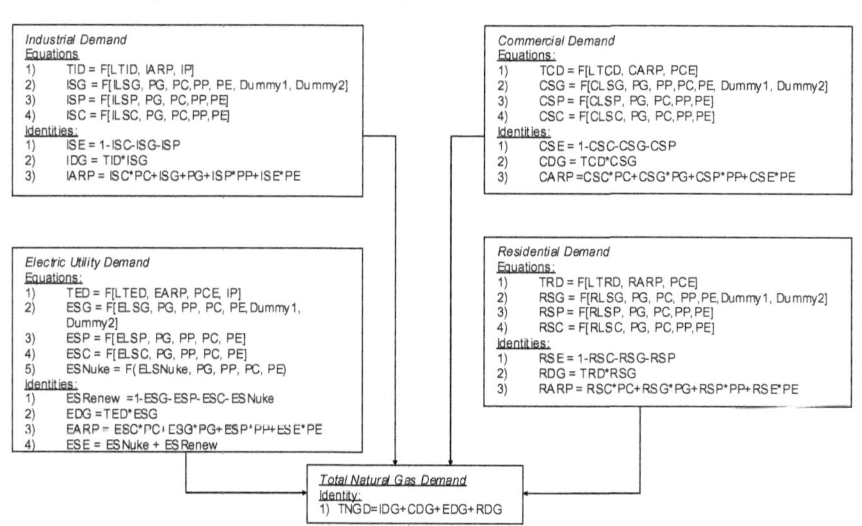

Variables are defined for the industrial sector, the definitions are the same for the other sectors, however the values are specific to each sector.

TID: Total Industrial Demand
IARP: Estimator for Average Industrial Real Price of All Fuels
ISC: Coals Share of Industrial Expenditure
ISE: Electricitys Share of Industrial Expend.
ISP: Petroleums Share of Industrial Expend.
IDG: Industrial Demand for Gas
LTID: Lagged Total Industrial Demand
ILSC: Lagged Value Share of Coal (Ind.)
ILSG: Lagged Value Share of Gas (Ind.)
ILSP: Lagged Value Share of Petroleum (Ind.)
ILSE: Lagged Value Share of Electricity (Ind.)

TNGD: Total Demand for Natural Gas
PC: Real Price of Coal
PG: Real Price of Gas
PP: Real Price of Petroleum
PE: Real Price of Electricity
IP: Industrial Production
PCE: Personal Consumption Expenditures
Dummy1: Dummy variable for years 1968-1978
Dummy2: Dummy variable for years 1979-1984

Equations: 17 Identities: 13

measured by industrial production or personal consumption expenditures, and on an index price that accounts for all four fuels. The share of total expenditures for each fuel in each sector is a function of its price, the prices for other fuels, and the previous period share. Because shares are percentages of expenditures, the share coefficient is positive if demand is inelastic and negative if demand is elastic.

Total gas demand across sectors is estimated in three stages. First, sectoral total demand is multiplied by the index of all-fuel prices to arrive at an estimated sectoral expenditure level. Second, sector expenditures are multiplied by gas share and divided by gas price to find sectoral gas demands. Finally, total gas demand equals the sum of the demands in these sectors.

6.1.2. *Estimated equations*

Total gas demand is taken to be represented by "marketed production" for the period 1967–2003. The highest level of demand was realized in the early 1970s, at 22.6 trillion cubic feet in 1972. Demand fell to its lowest level, 16.9 trillion cubic feet, in 1983 and again in 1986. Production at peak was out of reserves of 224 trillion cubic feet, and at the lowest level, it was out of reserves of 177 trillion cubic feet in 1983 and 165 trillion cubic feet in 1986. Rates-of-take from contracted reserve base were roughly the same, at one cubic foot per 10 cubic feet of reserve, in this period.

Prices paid by all types of consumers increased up to the mid-1980s, then held or declined until the mid-2000s. The industrial price in constant 1982 dollars doubled in the early 1970s, at a time when crude oil prices quadrupled during the OPEC crude embargo. In market terms, gas prices paid by industry were matching those of alternative fuels, which were escalating rapidly. Residential prices in constant 1982 dollars increased only 25 percent in this same period and wellhead prices increased by only the same dollar amount; both of these prices were regulated, whereas the industrial price was largely unregulated.

The gas take rate of both industrial and retail consumer declined in the early 1980s, however, and as a result, industrial and residential prices decreased. The constant dollar industrial price peaked in 1983 at $3.88 per Mcf to fall to $2.73 per Mcf in 1986, while the constant dollar residential price was $5.63 Mcf in 1983 and $4.92 Mcf in 1986. Demand growth was restored in the 1990s, and prices stabilized in this range, then increased to twice this range in the early 2000s.

These data for quantities and prices make up a set of annual observations for the period 1960–2003. Observations for each year for the determining variables were compiled so that 36 to 44 annual observations formed a data set for estimating demand equations, first for the regulatory and then the entire period.

The share equations are shown in Tables 5.2 through 5.5. They belong to a class of simultaneous equations fitted by Two-Stage Least Squares regression. In the first stage, values for the endogenous variables on the right-hand side of the equations were estimated from ordinary least squares regressions of these variables on all exogenous variables. Those endogenous variables are the price of gas (PG) and four sectional average energy prices (IARP, CARP, RARP and EARP). In the second stage, ordinary least squares regressions of the endogenous dependent variables are estimated on the exogenous and fitted endogenous variables from the first stage.

These equations trace the historical data closely. Both "old" equations with lagged dependent variables and "new" without lags are shown in these tables. Total sectional demands are determined by the energy price index and by either industrial or personal consumption (i.e., these independent variables have statistically significant coefficients). Gas shares are elastic in industrial and residential markets, and inelastic in commercial markets, but only the industrial price in the old equation is statistically significant.

Figure 6.2. Industrial Demand for Natural Gas

1. Total Industrial Energy Demand (TID)

Variable	Description	Old Estimation	New – No Lags
Intercept		4.954	20.80409
		(3.66)	(16.61)
LTID	Lagged Total	0.791	—
	Industrial Demand	(12.27)	—
IARP	Average Industrial	−0.536	−0.71974
	Energy Price	(−4.51)	(−2.50)
IP	Industrial Production	0.027	0.057955
		(3.19)	(2.87)

2. Industrial Share of Gas (ISG)

Variable	Description	Old Estimation	New – No Lags
Intercept		0.185	0.4663
		(3.56)	(10.06)
ILSG	Lagged Industrial	0.720	—
	Share of Gas	(11.40)	—
PG	Price of Gas	−0.018	−0.0122
		(−3.13)	(−1.04)
PC	Price of Coal	−0.034	−0.0376
		(−4.97)	(−2.38)
PP	Price of Petroleum	0.011	0.0126
		(4.90)	(2.77)

(*Continued*)

Figure 6.2. (*Continued*)

Variable	Description	Old Estimation	New – No Lags
PE	Price of Electricity	−0.002	−0.0079
		(−0.93)	(−2.00)
Dummy1	Dummy for	0.009	0.0213
	1968–1978	(1.74)	(1.98)
Dummy2	Dummy for	−0.004	−0.0068
	1979–1984	(−0.92)	(−0.68)

3. Industrial Share of Petroleum (ISP)

Variable	Description	Old Estimation	New – No Lags
Intercept		0.141	0.3547
		(3.11)	(11.57)
ILSP	Lagged Industrial	0.806	—
	Share of Petroleum	(10.00)	—
PG	Price of Gas	0.028	0.0525
		(4.22)	(4.51)
PC	Price of Coal	0.028	0.0448
		(6.03)	(4.62)
PP	Price of Petroleum	−0.009	−0.0089
		(4.76)	(−2.20)
PE	Price of Electricity	−0.003	−0.0037
		(−2.12)	(−1.42)

(*Continued*)

Figure 6.2. (*Continued*)

4. Industrial Share of Coal (ISC)

Variable	Description	Old Estimation	New – No Lags
Intercept		−0.027	−0.0624
		(−1.82)	(−1.01)
ILSC	Lagged Industrial Share of Coal	0.940	—
		(19.56)	—
PG	Price of Gas	−0.004	−0.1113
		(−0.49)	(−4.76)
PC	Price of Coal	0.001	−0.024
		(0.25)	(−1.24)
PP	Price of Petroleum	0.0004	0.0057
		(−0.17)	(0.70)
PE	Price of Electricity	0.003	0.0236
		(1.59)	(4.51)

Source: Database for regression developed from: *The Basic Petroleum Handbook* (American Petroleum Institute, various years), *Monthly Energy Review* (Department of Energy, Energy Information Administration), *Natural Gas Annual* (DOE, EIA).

Note: t-statistics for estimated coefficients are shown in parentheses. For the new model (no lags), t-statistics were compiled using heteroskedasticity and autocorrelation robust standard errors (see, for example, Andrews, D. W. K., "Consistent Covariance Matrix Estimation", *Econometrica*, 59, 817–858 (1991)). See MacAvoy (2000) Table 2.3, page 25 for technical notes on two stage regressions.

Figure 6.3. Commercial Demand for Natural Gas

1. Total Commercial Energy Demand (TCD)

Variable	Description	Old Estimation	New – No Lags
Intercept		0.558	2.5873
		(3.44)	(9.82)
LTCD	Lagged Total Commercial	0.860	—
	Demand	(14.27)	—
CARP	Average Commercial	−0.040	−0.0277
	Energy Price	(−1.97)	(−0.51)
PCE	Personal Consumption	0.0003	0.0015
	expenditures	(2.32)	(8.35)

2. Commercial Share of Gas (CSG)

Variable	Description	Old Estimation	New – No Lags
Intercept		0.366	0.7074
		(2.88)	(16.25)
CLSG	Lagged Commercial	0.516	—
	Share of Gas	(2.86)	—
PG	Price of Gas	0.014	0.0469
		(0.87)	(4.28)
PC	Price of Coal	0.004	0.0542
		(0.82)	(3.65)
PP	Price of Petroleum	0.029	0.0073
		(1.81)	(1.70)

(*Continued*)

Figure 6.3. (*Continued*)

Variable	Description	Old Estimation	New – No Lags
PE	Price of Electricity	−0.016	−0.0298
		(−2.59)	(−8.03)
Dummy1	Dummy for 1968–1978	−0.008	0.0015
		(−0.77)	(0.15)
Dummy2	Dummy for 1979–1984	0.004	−0.0024
		(0.38)	(−0.26)

3. Commercial Share of Petroleum (CSP)

Variable	Description	Old Estimation	New – No Lags
Intercept		−0.006	−0.0315
		(−0.28)	(−0.29)
CLSP	Lagged Commercial Share of Petroleum	0.992	—
		(22.35)	—
PG	Price of Gas	0.017	−0.1472
		(1.33)	(−3.57)
PC	Price of Coal	−0.008	0.0401
		(−2.29)	(1.17)

(*Continued*)

Figure 6.3. (*Continued*)

Variable	Description	Old Estimation	New – No Lags
PP	Price of Petroleum	0.008	0.0015
		(1.06)	(0.10)
PE	Price of Electricity	0.001	0.0256
		(0.44)	(2.78)

4. Commercial Share of Coal (CSC)

Variable	Description	Old Estimation	New – No Lags
Intercept		−0.013	0.2005
		(−1.14)	(−5.16)
CLSC	Lagged Commercial	0.859	—
	Share of Coal	(24.09)	—
PG	Price of Gas	−0.002	−0.0842
		(−0.38)	(−5.70)
PC	Price of Coal	0.0000	−0.0341
		(0.07)	(−2.77)
PP	Price of Petroleum	−0.0003	0.0024
		(0.11)	(0.48)
PE	Price of Electricity	0.001	0.0244
		(0.95)	(7.40)

Source: As in Figure 6.2; see also notes to Figure 6.2.

Figure 6.4. Residential Demand for Natural Gas

1. Total Residential Energy Demand (TRD)

Variable	Description	Old Estimation	New – No Lags
Intercept		1.503	7.0933
		(4.19)	(16.88)
LTRD	Lagged Total Residential	0.814	—
	Demand	(16.13)	—
IARP	Average Residential	−0.138	−0.2865
	Energy Price	(−4.19)	(−3.04)
PCE	Personal Consumption	0.0005	0.0017
	Expenditures	(3.77)	(5.71)

2. Residential Share of Gas (RSG)

Variable	Description	Old Estimation	New – No Lags
Intercept		0.250	0.5385
		(3.89)	(16.02)
RLSG	Lagged Residential	0.601	—
	Share of Gas	(5.23)	—
PG	Price of Gas	−0.007	−0.0054
		(−1.05)	(−0.64)
PC	Price of Coal	0.0002	−0.0085
		(0.08)	(−0.74)

(*Continued*)

Figure 6.4. (*Continued*)

Variable	Description	Old Estimation	New – No Lags
PP	Price of Petroleum	0.012	0.0006
		(1.35)	(0.19)
PE	Price of Electricity	−0.003	−0.0012
		(−1.80)	(−0.40)
Dummy1	Dummy for 1968–1978	0.013	0.0025
		(2.11)	(0.32)
Dummy2	Dummy for 1979–1984	0.019	0.0326
		(3.29)	(4.48)

3. Residential Share of Petroleum (RSP)

Variable	Description	Old Estimation	New – No Lags
Intercept		0.015	0.0952
		(0.80)	(1.17)
RLSP	Lagged Residential	0.957	—
	Share of Petroleum	(20.03)	—
PG	Price of Gas	0.017	−0.1227
		(1.57)	(−3.96)
PC	Price of Coal	−0.010	0.0677
		(−3.60)	(2.63)

(*Continued*)

Figure 6.4. (*Continued*)

Variable	Description	Old Estimation	New – No Lags
PP	Price of Petroleum	0.004	−0.0048
		(0.56)	(−0.45)
PE	Price of Electricity	0.001	0.0153
		(0.71)	(2.21)

4. Residential Share of Coal (RSC)

Variable	Description	Old Estimation	New – No Lags
Intercept		−0.004	−0.0662
		(−1.11)	(−4.85)
RLSC	Lagged Residential Share of Coal	0.852	—
		(38.57)	—
PG	Price of Gas	−0.001	−0.0292
		(0.61)	(−5.63)
PC	Price of Coal	0.0002	−0.0125
		(0.86)	(−2.89)
PP	Price of Petroleum	−0.0002	0.001
		(−0.37)	(0.58)
PE	Price of Electricity	0.0003	0.0083
		(0.94)	(7.19)

Source: As in Figure 6.2.

Figure 6.5. Electric Demand for Natural Gas

1. Total Electric Utility Demand (TED)

Variable	Description	Old Estimation	New – No Lags
Intercept		0.795	−5.1258
		(1.42)	(−6.34)
LTED	Lagged Total Electric	0.905	—
	Demand	(13.31)	—
IARP	Average Electric Energy	−0.091	1.076
	Price	(−0.79)	(4.90)
PCE	Personal Consumption	−0.001	0.0013
	Expenditures	(−1.13)	(0.49)
IP	Industrial Production	0.060	0.2363
		(1.91)	(3.16)

2. Electric Share of Gas (ESG)

Variable	Description	Old Estimation	New – No Lags
Intercept		0.005	−0.0545
		(0.14)	(−0.58)
ELSG	Lagged Electric Share of	0.885	—
	Gas	(12.41)	—
PG	Price of Gas	−0.022	−0.1245
		(−1.75)	(−5.27)
PC	Price of Coal	0.002	−0.0947
		(0.49)	(−2.95)

(*Continued*)

Figure 6.5. (*Continued*)

Variable	Description	Old Estimation	New – No Lags
PP	Price of Petroleum	−0.013	−0.0024
		(−0.95)	(−0.26)
PE	Price of Electricity	0.002	0.029
		(0.72)	(3.63)
Dummy1	Dummy for 1968–1978	0.0000	0.0752
		(0.003)	(3.44)
Dummy2	Dummy for 1979–1984	0.015	0.1237
		(1.40)	(6.07)

3. Electric Share of Petroleum (ESP)

Variable	Description	Old Estimation	New – No Lags
Intercept		0.072	0.2821
		(2.83)	(6.82)
ELSP	Lagged Electric	1.018	—
	Share of Petroleum	(10.42)	—
PG	Price of Gas	0.024	0.0048
		(3.17)	(0.31)
PC	Price of Coal	−0.011	0.1513
		(−4.14)	(11.58)
PP	Price of Petroleum	−0.001	−0.0046
		(−0.04)	(−0.84)
PE	Price of Electricity	−0.003	−0.0234
		(−1.33)	(−6.66)

(*Continued*)

Figure 6.5. (*Continued*)

4. Electric Share of Coal (ESC)

Variable	Description	Old Estimation	New – No Lags
Intercept		0.032	0.3059
		(0.60)	(7.32)
ELSC	Lagged Electric	0.773	—
	Share of Coal	(6.37)	—
PG	Price of Gas	−0.004	0.0037
		(−0.39)	(0.23)
PC	Price of Coal	0.004	−0.1261
		(0.98)	(−9.55)
PP	Price of Petroleum	−0.024	0.0014
		(−1.32)	(0.26)
PE	Price of Electricity	0.006	0.0222
		(2.18)	(6.26)

5. Electric Share of Nuclear (ESNuke)

Variable	Description	Old Estimation	New – No Lags
Intercept		0.017	0.2462
		(0.86)	(1.86)
ELSNuke	Lagged Electric	1.014	—
	Share of Nuclear	(33.44)	—

(*Continued*)

Figure 6.5. (*Continued*)

Variable	Description	Old Estimation	New – No Lags
PG	Price of Gas	0.010	0.1552
		(1.04)	(3.09)
PC	Price of Coal	−0.004	−0.0293
		(−1.31)	(−0.70)
PP	Price of Petroleum	0.013	−0.0143
		(2.09)	(−0.82)
PE	Price of Electricity	−0.001	−0.0143
		(−0.78)	(−1.27)

Source: as in Figure 6.2.

Event period variables are included to account for the early (NGPA) deregulatory impact on demand. Binary (0,1) variables for the "price control" years (1968–1978), and the "deregulation" years (1979–1984) of WGPA-phasing, have been included in the determinants of gas share in each sector. For the price control years, residential and commercial gas shares were lower and industrial share was higher. For the phased deregulation years, residential and commercial gas shares were higher, while industrial share was lower. The inference that the regulatory price control process at wholesale had built in subsides to industrial buyers seems to have been borne out by these share estimates.

6.2. Estimation of Supply Equations

The supply side of the model consists of six equations, as described in Figure 5.6. The first equation in the old model is for gas wells, which depends on previous period well drilling, gas and crude oil prices, and the relevant regulatory regime. In the new model, the number of wells depends only on current prices and regulatory regimes. This drilling activity then generates reserve discoveries and leads to revisions and extensions of old reserves.

The two equations for discoveries are for the number of new wells and new reserves per well. New wells are a function of previous period wells in the old model, gas and oil prices, and the presence of a specific regulatory regime. The average volume of reserves per new well is a function of gas price, new wells and the regulatory regime (with or without lagged discoveries).

The two equations for revision and extension of reserves have the same structure as the discoveries equations. The first determines the number of development wells, and the second determines the revision and extension of gas reserves per development well. The number of development wells depends on previous total and current exploratory wells, the gas price and in the old equation the number of previous period development wells. The regulatory regime in place at the time plays a role. Variables have been included for regulation on the hypothesis that the regulatory regime changes development activity.

Together, discoveries and revisions or extensions determine additions to reserves. The rate-of-take from reserves determines the amount of production. This rate in a function of total wells, industrial production and lagged rate-of-take. The sequence in which these relationships generate supply is shown in Figure 5.6.

Figure 6.6. Flow Diagram: Natural Gas Supply

Total Gas Wells
Equation:
1) TGW = F[LagTGW, SGP, SOP, Dummy1, Dummy2]

Discoveries
Equations:
1) NFW = F[LagNFW, SGP, SOP, Dummy1, Dummy2]
2) MND = F[LagMND, PG, TGW, CTGW, Dummy1, Dummy2]
Identity:
1) D = NFW*MND

Revisions & Extensions
Equations:
1) DEV = F[LagNFW, LagDEV, STGW, PG, Dummy1, Dummy2]
2) MNRE = F[LagMND, LagMNRE, TGW, Dummy1, Dummy2]
Identity:
1) RE = DEV*MNRE

RESERVES
Identity:
1) $R = LagR + D + RE - Q_p + Q_{Historic\ Difference}$

PRODUCTION
Equation:
1) PR = F[LagPR, STGW, IP, Dummy1, Dummy2]
Identity:
1) $Q_p = PR*R$
2) $Q_c = Q_p + Q_{Historic\ Excess}$

TGW: Total Gas Wells
STGW: % Change in TGW
CTGW: Cumulative TGW
LagTGW: Lagged Value of Total Gas Wells
SGP: % Change in Gas Price
SOP: % Change in Oil Price
NFW: New Field Wildcats
LagNFW: Lagged Value of New Field Wildcats
MND: New Discoveries per New Field Wildcat
LagMND: Lagged Value of Mean New Discoveries per New Field Wildcat
D: New Discoveries
DEV: Development Wells
LagDEV: Lagged Value of Developmental Wells

MNRE: Revisions & Extensions over Development Wells
RE: Revisions & Extensions
PR: Production-Reserve Ratio
LagPR: Lagged Value of Production-Reserve Ratio
Q_p: Production
Q_c: Consumption
IP: Industrial Production
R: Reserves
LagR: Lagged value of Reserves
$Q_{Historic\ Excess}$: Historic Excess Demand
$Q_{Historic\ Difference}$: Historic balancing item
Dummy1: Dummy variable for years 1968-1978
Dummy2: Dummy variable for years 1979-1984

Equations: 6 Identities: 4

The regression equations for supply fitted to the old and new price and quantity data are shown in Figure 6.7. The first equation, for total gas wells indicates that increases in the number of wells have been associated with changes in gas and oil price (although the coefficients of the price variables are not statistically significant in either old or new equations). Previous period well installation rates were positive and significant in year-to-year levels of total wells.

Once the number of wells is determined, the equation set separates activity between discovery wells to locate new reserves and development wells to add further to old reserves. Discovery wells have been determined by three variables — changes in oil and gas prices, and lagged wells, all of which have positive and significant coefficients in the old equation. This discovery well completion rate has been determined by the gas price, which has a positive and significant coefficient, the change in total wells, and lagged development wells in the new equation, all with positive and significant coefficients. The one-period lagged rate of completion of these wells has a positive, but insignificant effect on the number of new development wells. That the accumulation of new reserves has been a two-stage process of discovery and development is borne out by these fitted equations.

As the process of adding reserves takes place, both the amount found in new discoveries declines and the amount added in extensions of old reserves declines. In effect, diminishing returns at any field location set in. The size of a new discovery decreases (i.e., the coefficient of the total wells variable in the mean new discovery equation is negative) and the coefficient for the lagged size of discoveries is less than one. The size of extension and revisions of reserves decrease with total wells to also bring about diminishing returns.

Additions to reserves in a given year have been the product of number of wells times additions per well. That is, discovered reserves equal new field wells times discoveries per well, and reserve revisions and extensions equal development wells times revisions per well. The sum equals new reserves, to then provide additional production. Gas supply equals the rate of the production out of these reserves. The equations, as shown in Table 6.7, have significant coefficients for industrial production, the change in total wells, and the lagged rate-of-take (in the old equations). Production in a given year equals current reserves multiplied by this rate of production from reserves.

The regulatory regime has played a role as a determinant of supply. The coefficient for the binary variable for the 1968–1978 period of gas price caps shows that the production-reserve ratio increased in that period. With price limits in place, the regime reduced discoveries, extensions and revisions but these increase in the new models for the years immediately following the Natural Gas Policy Act (NGPA).

The equations together describe the consumer purchase level of demand and the wellhead level of supply. Market behavior is affected by regulatory activity that centers on the late 1960s to middle 1970s, when wellhead price limits were set in place by the Federal Power Commission, and on the period from the late 1970s to mid-1980s, when the Federal Energy Regulatory Commission implemented phased price decontrol under the Natural Gas Policy Act of 1978.

The more recent responses of industry to partial deregulation can be indicated by model simulation. The supply and demand system can be solved for production equal to demand, while setting the regulatory variables for Order 636 at zero so as to specify partially "market-determined" price.

Figure 6.8 and 6.9 compare predicted and actual demand and supply of natural gas. The first shows historic series for consumption and production, and the second shows predictions from the new model without lags using observations from 1961 to 2001. Also, excess demand as simulated using old and new model estimates is reported here. The series shown in Figure 6.9 are not equilibrium quantities since they were computed using actual prices.

Figure 6.7. Natural Gas Supply

1. Total Gas Wells (TGW)

Variable	Description	Old Estimation	New – No Lags
Intercept		1,066.380	7,646.093
		(0.131)	(8.85)
SGP	Change in Gas Price	5,564.133	752.2223
		(1.69)	(0.13)
SOP	Change in Oil Price	2,989.180	−120.532
		(1.10)	(−0.12)
LagTGW	Lagged Total Gas Wells	0.832	—
		(9.20)	—
Dummy1	Dummy for 1968—1978	190.783	−472.734
		(0.26)	(−0.33)
Dummy2	Dummy for 1979—1984	1,235.313	9,398.98
		(1.05)	(4.72)

2. New Field Wells (NFW)

Variable	Description	Old Estimation	New – No Lags
Intercept		147.691	715.2065
		(1.73)	(7.67)
SGP	Change in Gas Price	754.476	590.7696
		(1.94)	(0.93)

(*Continued*)

Figure 6.7. (*Continued*)

Variable	Description	Old Estimation	New – No Lags
SOP	Change in Oil Price	476.555	28.40332
		(1.57)	(0.26)
LagNFW	Lagged New Field	0.762	—
	Wells	(7.86)	—
Dummy1	Dummy for	69.817	234.1213
	1968–1978	(0.87)	(1.52)
Dummy2	Dummy for	169.486	1,153.887
	1979–1984	(1.15)	(5.37)

3. Mean New Discoveries (MND)

Variable	Description	Old Estimation	New – No Lags
Intercept		—	0.013933
		—	(9.56)
PG	Gas Price	−0.001	−0.00064
		(−0.47)	(−0.35)
TGW	Total Gas Wells	0.0001	0.0000
		(0.60)	(−2.15)
CTGW	Cumulative Total Gas	0.0000	−0.0000
	Wells	(0.60)	(−1.15)
LagMND	Lagged mean New	0.863	—
	Discoveries	(8.56)	—

(*Continued*)

Figure 6.7. (*Continued*)

Variable	Description	Old Estimation	New – No Lags
Dummy1	Dummy for 1968–1978	−0.001 (−0.90)	−0.00288 (−2.94)
Dummy2	Dummy for 1979–1984	0.0012 (0.66)	0.000858 (0.46)

4. Development Wells (DEV)

Variable	Description	Old Estimation	New – No Lags
Intercept		463.629 (0.83)	822.3112 (1.74)
PG	Gas Price	2,346.686 (2.21)	4,606.089 (9.82)
STGW	Change in Total Gas Wells	10,070.622 (5.62)	0.676008 (6.11)
LagNFW	Lagged New Field Wells	1.250 (1.45)	2.44173 (3.91)
LagDEV	Lagged Development Wells	0.456 (2.25)	— —
Dummy1	Dummy for 1968–1978	−409.111 (−0.73)	−308.361 (−0.65)
Dummy2	Dummy for 1979–1984	227.338 (0.24)	−511.987 (−0.62)

(*Continued*)

Figure 6.7. (*Continued*)

5. Mean New Revisions and Extenstions (MNRE)

Variable	Description	Old Estimation	New – No Lags
Intercept		—	0.002619
		—	(2.84)
TGW	Total Gas Wells	0.0000	−0.0000
		(1.63)	(−2.42)
LagMND	Lagged Mean	0.110	0.14102
	New Discoveries	(2.23)	(2.09)
LagMNRE	Lagged Mean New	0.625	—
	Revisions and	(4.61)	—
	Extenstions		
Dummy1	Dummy for	−0.001	−0.00113
	1968–1978	(−3.35)	(−3.98)
Dummy2	Dummy for	−0.001	0.000376
	1979-1984	(−179)	(0.64)

6. Production Reserve Ratio (PR)

Variable	Description	Old Estimation	New – No Lags
Intercept		0.005	0.009538
		(2.83)	(2.2)
IP	Industrial Production	0.0003	0.000913
		(5.06)	(18.88)

(*Continued*)

Figure 6.7. (*Continued*)

Variable	Description	Old Estimation	New – No Lags
STGW	Change in Total Gas Wells	0.019 (5.12)	1.15E-06 (1.82)
LagPR	Lagged Production Reserve Ratio	0.668 (10.63)	— —
Dummy1	Dummy for 1968–1978	0.003 (2.47)	0.011814 (4.76)
Dummy2	Dummy for 1979–1984	0.003 (1.57)	0.015377 (5.73)

Source: as in Figure 6.2.

As could be expected, the new model, based on estimates using all available data, produces in general more accurate forecasts. However, comparisons should be made with care: the new results are forecasts inside the sample period used for model estimation. Old model forecasts are out of sample predictions and by their nature should be less accurate. But according to the new model results there should have been excess demand for gas in 2000 and 2002 which are in line with the gas "shortages" that did occur for that second year.

The new model provides more accurate forecasts of the equilibrium price of gas. It is still incapable of predicting

Figure 6.8. Historic Demand and Supply of Natural Gas

Year	Historic Price of Gas	Historic Consumption (Demand)	Historic Production (Dry) and Net Imports	Excess Demand
1997	1.47	22.16	21.19	0.98
1998	1.22	21.57	21.37	0.20
1999	1.35	21.81	21.69	0.12
2000	2.23	22.89	21.98	0.90
2001	2.43	22.07	22.44	−0.37
2002	1.72	21.87	21.99	−0.12

Source: As explained in the text.

large changes from demand shocks, but it captures the long-run trend while the old model does not. In this sense, then, the elimination of long term contracts and of price controls at the wellhead, with partial decontrol of transport prices, shifted demand functions to higher levels, increased reserves and production, and allowed markets to come closer to clearing at higher prices.

Forecasts of how the system will operate under the new conditions are displayed in Figure 6.10 for a natural gas market for the period 2002–2010.

These predictions differ from those from the old model–there is no downward trend for prices, and the new model

Figure 6.9. Predicted Excess Demand for Natural Gas at Historic Prices

Year	Historic Price of Gas	Predicted Demand	Predicted Supply	Predicted Excess Demand New Estimates (1961–2001)	Predicted Excess Demand Old Estimates (1961–1994)
1997	1.47	22.58	21.62	0.96	0.08
1998	1.22	23.16	23.01	0.15	1.91
1999	1.35	24.06	23.28	0.78	0.47
2000	2.23	23.51	22.02	1.48	−4.31
2001	2.43	22.16	22.57	−0.41	−4.17
2002	1.72	24.03	21.92	2.11	−1.98

Source: As explained in the text.

Figure 6.10. New Model Simulated Prices, Production and Reserves of Natural Gas

	Current Gas Market		
	Price	Production	Gas Reserves
2003	1.38	21.72	168.68
2004	1.49	23.70	171.52
2005	1.68	24.30	169.26
2006	2.11	23.42	163.68
2007	1.78	24.71	152.29
2008	1.77	23.74	159.75
2009	1.88	24.94	159.92
2010	1.84	25.89	157.32

Source: As explained in the text. Price in $ per Mcf, production and reserves in trillions of cubic feet.

predicts higher average prices and a price spike in 2006. Predicted production continues to increase at historic rates but reserve forecasts are much lower than historic levels. In general, the deregulation model predicts higher prices and lower reserves.

Index